MATH PROOFS DEMYSTIFIED

Demystified Series

Advanced Statistics Demystified
Algebra Demystified
Anatomy Demystified
Astronomy Demystified
Biology Demystified
Biotechnology Demystified
Business Statistics Demystified
Calculus Demystified
Chemistry Demystified
College Algebra Demystified
Differential Equations Demystified
Digital Electronics Demystified
Earth Science Demystified
Electricity Demystified
Electronics Demystified
Environmental Science Demystified
Everyday Math Demystified
Geometry Demystified
Math Proofs Demystified
Math Word Problems Demystified
Microbiology Demystified
Physics Demystified
Physiology Demystified
Pre-Algebra Demystified
Precalculus Demystified
Probability Demystified
Project Management Demystified
Quantum Mechanics Demystified
Relativity Demystified
Robotics Demystified
Statistics Demystified
Trigonometry Demystified

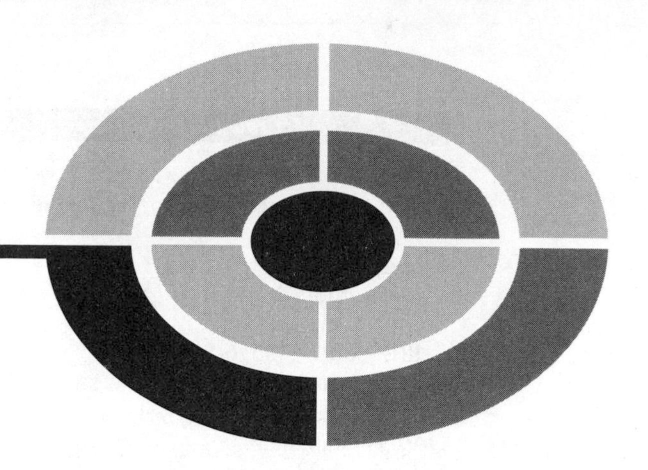

MATH PROOFS DEMYSTIFIED

STAN GIBILISCO

McGRAW-HILL
New York Chicago San Francisco Lisbon London
Madrid Mexico City Milan New Delhi San Juan
Seoul Singapore Sydney Toronto

The McGraw-Hill Companies

Library of Congress Cataloging-in-Publication Data

Gibilisco, Stan.
Math proofs demystified / Stan Gibilisco.
p. cm.
ISBN 0-07-144576-5
1. Proof theory—Popular works. I. Title.

QA9.54.G53 2005
511.3′6—dc22

2005041681

1 2 3 4 5 6 7 8 9 0 DOC/DOC 0 1 0 9 8 7 6 5

ISBN 0-07-144576-5

The sponsoring editor for this book was Judy Bass and the production supervisor was Pamela A. Pelton. It was set in Times Roman by Fine Composition. The art director for the cover was Margaret Webster-Shapiro; the cover designer was Handel Low.

Printed and bound by RR Donnelley.

This book is printed on recycled, acid-free paper containing a minimum of 50% recycled, de-inked fiber.

To Samuel, Tim, and Tony from Uncle Stan

CONTENTS

FOREWORD

This book deals with the idea and practice of proof in mathematics. As a college teacher, I know that this is a difficult concept to grasp, and a major poser for both teachers and learners. As a Gibilisco reader, I wasn't expecting anything less than a complete, entertaining, and go-getting presentation. I have been amply rewarded in my expectations.

Chapter 1 gets you right in the midst of the symbols that enable you to read a mathematical argument. You need this, just as a music student needs to know how to read a score. Chapter 2 deals with more sophisticated logic: how to put thoughts together coherently (and correctly—your typical mathematician is not a politician). Chapter 3, now that you have the language, actually builds a mathematical universe; in this it is a visionary chapter, yet it feels natural, and it is beautifully done. In Chapter 4, the fun begins! The mind-bending problems of fallacies and paradoxes are well illustrated. Chapters 5 and 6 are a bit more traditional, and provide an excellent selection of basic facts in geometry and numbers, respectively. Chapter 7 concludes the book with an innovative and mind-opening overview of some famous proofs. This can be read even "if only" to learn about, and savor, the development of mathematics in history as an intellectual adventure.

The book can be used for self-training. It assumes nothing, and teaches you everything you need. How it teaches you is another story. Stan Gibilisco has the gift and the passion of a coach. He provides the right example and exercise as soon as you see something new; by going through it with him, and again on your own in the quiz at the end of each chapter, you make it your own. Gibilisco takes you there, and is with you each step of the way.

When Stan Gibilisco asked me to write a short foreword for this book, I felt honored. I knew, in this case, that he wanted to distance himself from the material for two reasons. First, he has a personal attachment to proofs. (I've seen a mathematical journal that Stan kept as a college student, where he challenged himself to create an alternative concept of number and function, to supply some of the properties that the theorems he was taught did not have. He came close to doing something like what Bernhard Riemann did in the nineteenth century when he created the concept of a Riemann surface.) The second reason why Stan asked someone else to write about the book is, I think, that he was not complacent. He had decided to undertake a formidable task: portray the very language of mathematics. Stan wanted to provide the basics and a little more, a true exposure to the curiosity and creativity that has driven people, through the ages, to attempt to envision all possible worlds. It was to be a friendly book—as are all in the *Demystified* series—and also an abstract work that would show you beautiful examples and help you to soar high towards truth. Its reader-friendliness is of a sort that Gibilisco's readers have come to know. Its beauty must reside in the mind of the audience. As the Indian mathematician Bhaskara II said in the 12th century, "Behold!" (That was his proof-without-words of the Theorem of Pythagoras, which is illustrated in Chapter 7 of this book.)

Please enjoy this book and keep it handy! If I see you in my Algebra class, I will know you from it.

EMMA PREVIATO, Professor of Mathematics
Department of Mathematics and Statistics, Boston University

PREFACE

This book is for people who want to learn how to prove mathematical theorems. It can serve as a supplemental text in a classroom, tutored, or home-schooling environment. It should also be useful for people who need to refresh their knowledge of, or skills at, this daunting aspect of mathematics.

For advancing math students, the introduction to theorem-proving can be a strange experience. It is more of an art than a science. In many curricula, students get their first taste of this art in middle school or high school geometry. I suspect that geometry is favored as the "launching pad" for theorem-proving because this field lends itself to concrete illustrations, which can help the student see how proofs progress. This book starts out at a more basic level, dealing with the principles of "raw logic" before venturing into any specialized field of mathematics.

This book contains practice quizzes, tests, and exam questions. In format, they resemble the questions found in standardized tests. There is a short quiz at the end of each chapter. These quizzes are all "open book." You may (and should) refer to the chapter texts when taking them. This book has two multi-chapter sections or "parts," each of which concludes with a test. Take each test when you're done with all the chapters in the applicable section. There is a "closed book" exam at the end of this course. It contains questions drawn uniformly from all the chapters. Take it when you have finished both sections, both section tests, and all the quizzes.

In the back of the book, there is an answer key for all the quizzes, both tests, and the final exam. Each time you've finished a quiz, test, or the exam, have a friend check your paper against the answer key and tell you your score without letting you know which questions you missed. Keep studying until you can get at least three-quarters (but hopefully nine-tenths) of the answers right.

As I wrote this work, I tried to strike a balance between the "absolute rigour" that G. H. Hardy demanded in the early 1900s when corresponding with Ramanujan, the emerging Indian number theorist, and the informality that tempts everybody who tries to prove anything. I decided to employ a conversational style in a field where some purists will say that such language is out of place. It was my desire to bridge what sometimes seemed like an intellectual gulf that couldn't be spanned by any author. I hope the result is a course that will, at least, leave serious students better off after completing it than they were before they started.

Some college and university professors are concerned that American math students aren't getting enough training in logic and theorem-proving at the middle school and high school levels. These skills are essential if one is to develop anything new in mathematics. Sound reasoning is mandatory if one hopes to become a good theoretical scientist, experimentalist, or engineer—or even a good trial lawyer.

I recommend that you complete one chapter every couple of weeks. That will make the course last approximately one standard semester. Two hours a day ought to be enough study time. I also recommend you read as many of the "Suggested Additional References" (listed in the back of this book) as you can. Dare I insinuate that mathematics can be *cool*?

Illustrations in this book were generated with CorelDRAW. Some of the clip art is courtesy of Corel Corporation.

Suggestions for future editions are welcome.

STAN GIBILISCO

ACKNOWLEDGMENTS

I extend heartfelt thanks to Emma Previato, Professor of Mathematics at Boston University, and Bonnie Northey, a math teacher and good friend, who helped me with the proofreading of the manuscript for this book. I also thank my nephew Tony Boutelle, a student at Macalester College in St. Paul, for taking the time to read the manuscript and offer his insight from the point of view of the intended audience.

MATH PROOFS DEMYSTIFIED

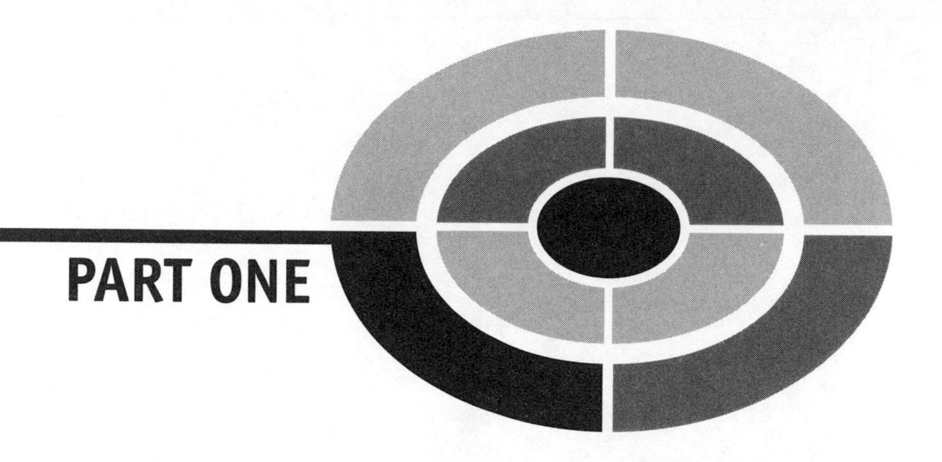

PART ONE

The Rules of Reason

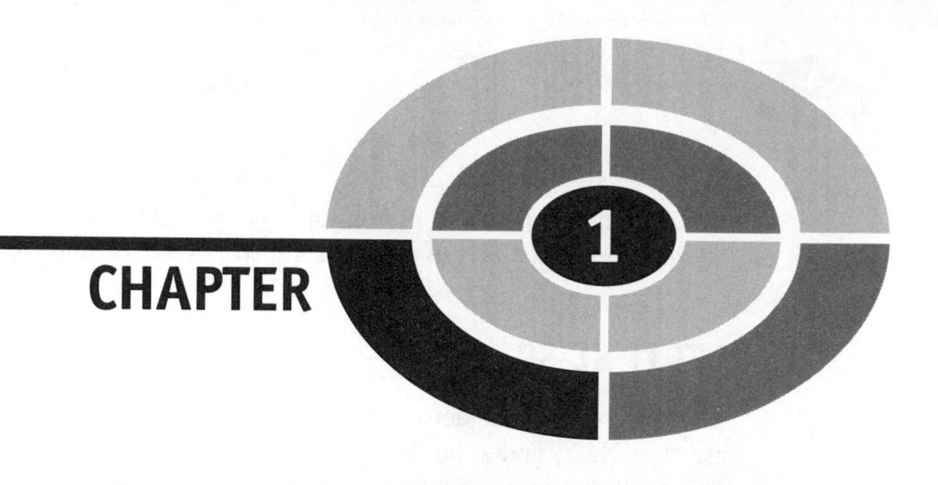

The Basics of Propositional Logic

In order to prove something, we need a formal system of reasoning. It isn't good enough to have "a notion" or even "a powerful feeling" that something is true or false. We aren't trying to convince a jury that something is true "beyond a reasonable doubt." In mathematics, we must be prepared to demonstrate the truth of a claim so there is no doubt whatsoever.

To understand how proofs work, and to learn how to perform them, we must become familiar with the laws that govern formal reasoning. *Propositional logic* is the simplest scheme used for this purpose. It's the sort of stuff Socrates taught in ancient Greece. This system of logic is also known as *sentential logic*, *propositional calculus*, or *sentential calculus*.

Operations and Symbols

The word *calculus* in logic doesn't refer to the math system invented by Newton and Leibniz that involves rates of change and areas under curves. In logic, *calculus* means a formal system of reasoning. The words *propositional* or *sentential* refer to the fact that the system works with complete sentences.

LET IT BE SO!

You will often come across statements in math texts, including this book, such as: "Let X, Y, and Z be logical variables." This language is customary. You'll find it all the time in mathematical literature. When you are told to "let" things be a certain way, you are being asked to imagine, or suppose, that things are that way. This sets the scene in your mind for statements or problems to follow.

SENTENCES

Propositional logic does not involve breaking sentences down into their internal details. We don't have to worry about how words are interconnected and how they affect each other within a sentence. Those weird sentence diagrams, which you may have worked with in your middle-school grammar class, are not a part of propositional logic. A *sentence*, also called a *proposition*, is the smallest possible entity in propositional logic.

Sentences are represented by uppercase letters of the alphabet. You might say "It is raining outside," and represent this by the letter R. Someone else might add, "It's cold outside," and represent this by the letter C. A third person might say, "The weather forecast calls for snow tomorrow," and represent this by the letter S. Still another person might add, "Tomorrow's forecast calls for sunny weather," and represent this by B (for "bright"; we've already used S).

NEGATION (NOT)

When we write down a letter to stand for a sentence, we assert that the sentence is true. So, for example, if John writes down C in the above situation, he means to say "It is cold outside." You might disagree if you grew up in Alaska and John grew up in Hawaii. You might say, "It's not cold outside." This can be symbolized as the letter C with a *negation* symbol in front of it.

There are several ways in which negation, also called *NOT*, can be symbolized. In propositional logic, a common symbol is a drooping minus sign (¬). That's the one we'll use. Some texts use a tilde (~) to represent negation. Some use a minus sign (–). Some put a line over the letter representing the sentence; still others use an accent symbol. It seems as if there is no shortage of ways to express a denial, even in symbolic logic! In our system, the sentence "It's not cold outside" can be denoted as ¬C.

Suppose someone comes along and says, "You are correct to say ¬C. In fact, I'd say it's hot outside!" Suppose this is symbolized by H. Does H mean the same thing as ¬C? Not necessarily. You've seen days that were neither cold nor hot. There can be in-between states such as "cool" (K), "mild" (M), and "warm" (W). But there is no in-between condition when it comes to C and ¬C. In propositional logic, either it is cold, or else it is not cold. Either it is hot, or else it is not hot. A proposition is either true, or else it is false (not true).

There are logical systems in which in-between states exist. These go by names such as *fuzzy logic*. But a discussion of those types of logic belongs in a different book. In all the mathematical proofs we'll be dealing with, any proposition is either true or false; there is neither a neutral truth state nor any continuum of truth values. Our job, when it comes to doing math proofs, is to demonstrate truth or falsity if we can.

CONJUNCTION (AND)

Propositional logic doesn't get involved with how the phrases inside a sentence affect each other, but it is very concerned with the ways in which distinct, complete sentences interact in logical discourse. Sentences can be combined to make bigger ones, called *compound sentences*. The truth or falsity of a compound sentence depends on the truth or falsity of its components, and on the ways those components are connected.

Suppose someone says, "It's cold outside, and it's raining outside." Using the symbols from the previous sections, we can write this as:

C AND R

In logic, we use a symbol in place of the word AND. There are several symbols in common use, including the ampersand (&), the inverted wedge (∧), the asterisk (*), the period (.), the multiplication sign (×), and the raised dot (·). We'll use the ampersand because it represents the word AND in everyday language, and is easiest to remember. Thus, the compound sentence becomes:

C & R

The formal term for the AND operation is *logical conjunction*. A compound sentence containing one or more conjunctions is true when, but only when, both or all of its components are true. If any of the components are false, then the whole compound sentence is false.

DISJUNCTION (OR)

Now imagine that a friend comes along and says, "You are correct in your observations about the weather. It's cold and raining; there is no doubt about those facts. I have been listening to the radio, and I heard the weather forecast for tomorrow. It's supposed to be colder tomorrow than it is today. But it's going to stay wet. So it might snow tomorrow."

You say, "It will rain or it will snow tomorrow, depending on the temperature."

Your friend says, "It might be a mix of rain and snow together, if the temperature is near freezing."

"So we might get rain, we might get snow, and we might get both," you say.

"Correct. But the weather experts say we are certain to get precipitation of some sort," your friend says. "Water is going to fall from the sky tomorrow—maybe liquid, maybe solid, and maybe both."

In this case, suppose we let R represent the sentence "It will rain tomorrow," and we let S represent the sentence "It will snow tomorrow." Then we can say:

S OR R

This is an example of *logical disjunction*. There are at least two symbols commonly used to represent disjunction: the addition symbol (+) and the wedge (∨). Let's use the wedge. We can now write:

$$S \vee R$$

A compound sentence in which both, or all, of the components are joined by disjunctions is true when, but only when, at least one of the components is true. A compound sentence made up of disjunctions is false when, but only when, all the components are false.

Logical disjunction, as we define it here, is the *inclusive OR* operation. There's another logic operation called *exclusive OR*, in which the compound sentence is false, not true, if and only if all the components are true. We won't deal with that now. The *exclusive OR* operation, sometimes abbreviated XOR, is important when logic is applied in engineering, especially in digital electronic circuit design.

IMPLICATION (IF/THEN)

Imagine that the conversation about the weather continues. You and your friend are trying to figure out if you should get ready for a snowy day tomorrow, or whether rain and gloom is all you'll have to contend with.

"Does the weather forecast say anything about snow?" you ask.

"Not exactly," your friend says. "The radio announcer said, 'There is going to be precipitation through tomorrow night, and it's going to get colder tomorrow.' I looked at my car thermometer as she said that, and it said the outdoor temperature was just a little bit above freezing."

"If there is precipitation, and if it gets colder, then it will snow," you say.

"Of course."

"Unless we get an ice storm."

"That won't happen."

"Okay," you say. "If there is precipitation tomorrow, and if it is colder tomorrow than it is today, then it will snow tomorrow." (This is a weird way to talk, but we're learning about logic, not the art of witty conversation.)

Suppose you use P to represent the sentence "There will be precipitation tomorrow." In addition, let S represent the sentence "It will snow tomorrow," and let C represent the sentence "It will be colder tomorrow." Then in the previous conversation, you have made a compound proposition consisting of three sentences, like this:

IF (P AND C), THEN S

Another way to write this is:

(P AND C) IMPLIES S

In this context, the meaning of the term "implies" is intended in the strongest possible sense. In logic, if X "implies" Y, it means that X is always accompanied or followed by Y, not merely that X suggests Y. Symbolically, the above proposition is written this way:

$$(P \& C) \Rightarrow S$$

The double-shafted arrow pointing to the right represents *logical implication*, also known as the *IF/THEN operation*. In a logical implication, the "implying" sentence (to the left of the double-shafted arrow) is called the *antecedent*. In the previous example, the antecedent is (P & C). The "implied" sentence (to the right of the double-shafted arrow) is called the *consequent*. In this example, the consequent is S.

Some texts make use of other symbols for logical implication, including the "hook" or "lazy U opening to the left" ($\supset$), three dots ($\therefore$), and a single-shafted arrow pointing to the right ($\rightarrow$). In this book, we'll stick with the double-shafted arrow pointing to the right.

LOGICAL EQUIVALENCE (IFF)

Suppose your friend changes the subject and says, "If it snows tomorrow, then there will be precipitation and it will be colder."

For a moment you hesitate, because this isn't the way you'd usually think about this kind of situation. But you have to agree, "That is true. It sounds strange, but it's true." Your friend has just made this implication:

$$S \Rightarrow (P \;\&\; C)$$

Implication holds in both directions here, but there are plenty of scenarios in which an implication holds in one direction but not the other.

You and your friend have agreed that both of the following implications are valid:

$$(P \;\&\; C) \Rightarrow S$$
$$S \Rightarrow (P \;\&\; C)$$

These two implications can be combined into a conjunction, because we are asserting them both at the same time:

$$[(P \;\&\; C) \Rightarrow S] \;\&\; [S \Rightarrow (P \;\&\; C)]$$

When an implication is valid in both directions, the situation is defined as a case of *logical equivalence*. The above statement can be shortened to:

$$(P \;\&\; C) \text{ IF AND ONLY IF } S$$

Mathematicians sometimes shorten the phrase "if and only if" to the single word "iff." So we can also write:

$$(P \;\&\; C) \text{ IFF } S$$

The symbol for logical equivalence is a double-shafted, double-headed arrow ($\Leftrightarrow$). There are other symbols that can be used. Sometimes you'll see an equals sign, a three-barred equals sign ($\equiv$), or a single-shafted, double-headed arrow ($\leftrightarrow$). We'll use the double-shafted, double-headed arrow to symbolize logical equivalence. Symbolically, then:

$$(P \;\&\; C) \Leftrightarrow S$$

PROBLEM 1-1

Give an example of a situation in which logical implication holds in one direction but not in the other.

SOLUTION 1-1

Consider this statement: "If it is overcast, then there are clouds in the sky." This statement is true. Suppose we let O represent "It is overcast" and K represent "There are clouds in the sky." Then we have this, symbolically:

$$O \Rightarrow K$$

If we reverse this, we get a statement that isn't necessarily true. Consider:

$$K \Rightarrow O$$

This translates to: "If there are clouds in the sky, then it's overcast." We have all seen days or nights in which there were clouds in the sky, but there were clear spots too, so it was not overcast.

Truth Tables

The outcome, or *logic value*, of an operation in propositional logic is always either true or false, as we've seen. Truth can be symbolized as T, +, or 1, while falsity can be abbreviated as F, –, or 0. Let's use T and F. They are easy to remember: "T" stands for "true" and "F" stands for "false"! When performing logic operations, sentences that can attain either T or F logic values (depending on the circumstances) are called *variables*.

A *truth table* is a method of denoting all possible combinations of truth values for the variables in a proposition. The values for the individual variables, with all possible permutations, are shown in vertical columns at the left. The truth values for compound sentences, as they are built up from the single-variable (or *atomic*) propositions, are shown in horizontal rows.

TRUTH TABLE FOR NEGATION

The simplest truth table is the one for negation, which operates on a single variable. Table 1-1 shows how this works for a single variable called X.

Table 1-1. Truth Table for Negation

X	¬X
F	T
T	F

Table 1-2. Truth Table for Conjunction

X	Y	X & Y
F	F	F
F	T	F
T	F	F
T	T	T

TABLE FOR CONJUNCTION

Let X and Y be two logical variables. Conjunction (X & Y) produces results as shown in Table 1-2. The AND operation has value T when, but only when, both variables have value T. Otherwise, the operation has value F.

TABLE FOR DISJUNCTION

Logical disjunction for two variables (X ∨ Y) has a truth table that looks like Table 1-3. The OR operation has value T when either or both of the variables have value T. If both of the variables have value F, then the operation has value F.

Table 1-3. Truth Table for Disjunction

X	Y	X ∨ Y
F	F	F
F	T	T
T	F	T
T	T	T

Table 1-4. Truth Table for Implication

X	Y	$X \Rightarrow Y$
F	F	T
F	T	T
T	F	F
T	T	T

TABLE FOR IMPLICATION

A logical implication is valid (that is, it has truth value T) except when the antecedent has value T and the consequent has value F. Table 1-4 shows the truth values for logical implication.

PROBLEM 1-2

Give an example of a logical implication that is obviously invalid.

SOLUTION 1-2

Let X represent the sentence, "The wind is blowing." Let Y represent the sentence, "A hurricane is coming." Consider this sentence:

$$X \Rightarrow Y$$

Now imagine that it is a windy day. Therefore, variable X has truth value T. But suppose you are in North Dakota, where there are never any hurricanes. Sentence Y has truth value F. Therefore, the statement "If the wind is blowing, then a hurricane is coming" is false.

TABLE FOR LOGICAL EQUIVALENCE

If X and Y are logical variables, then X IFF Y has truth value T when both variables have value T, or when both variables have value F. If the truth values of X and Y are different, then X IFF Y has truth value F. This is broken down fully in Table 1-5.

Table 1-5. Truth Table for Logical Equivalence

X	Y	$X \Leftrightarrow Y$
F	F	T
F	T	F
T	F	F
T	T	T

THE EQUALS SIGN

In logic, we can use an ordinary equals sign to indicate truth value. Thus if we want to say that a particular sentence K is true, we can write K = T. If we want to say that a variable X always has false truth value, we can write X = F. Just be careful about this. Don't confuse the meaning of the equals sign with the meaning of the double-shafted, double-headed arrow that stands for logical equivalence!

PROBLEM 1-3

Tables 1-1 through 1-4—the truth tables for negation, conjunction, disjunction, and implication—are defined by convention. The truth values are based on common sense. Arguably, the same is true for logical equivalence. It makes sense that two logically equivalent statements ought to have identical truth values, and that if they don't, they can't be logically equivalent. But suppose you want to prove this. You can derive the truth values for logical equivalence based on the truth tables for conjunction and implication. Do it, and show the derivation in the form of a truth table.

SOLUTION 1-3

Remember that $X \Leftrightarrow Y$ means the same thing as $(X \Rightarrow Y)$ & $(Y \Rightarrow X)$. You can build up $X \Leftrightarrow Y$ in steps, as shown in Table 1-6 as you go from left to right. The four possible combinations of truth values for sentences X and Y are shown in the first (left-most) and second columns. The truth values for $X \Rightarrow Y$ are shown in the third column, and the truth values for $Y \Rightarrow X$ are shown in the fourth column. In order to get the truth values for the fifth (right-most) column, conjunction is applied to the truth values in the third and fourth columns. The *complex logical operation* (also called a *compound logical operation*

Table 1-6. Truth Table for Problem 1-3

X	Y	X ⇒ Y	Y ⇒ X	(X ⇒ Y) & (Y ⇒ X) which is the same as X ⇔ Y
F	F	T	T	T
F	T	T	F	F
T	F	F	T	F
T	T	T	T	T

because it's made up of combinations of the basic ones) in the fifth column is the same thing as X ⇔ Y.

Q.E.D.

What you have just seen is a mathematical proof of the fact that for any two logical sentences X and Y, the value of X ⇔ Y is equal to T when X and Y have the same truth value, and the value of X ⇔ Y is equal to F when X and Y have different truth values. Sometimes, when mathematicians finish proofs, they write "Q.E.D." at the end. This is an abbreviation of the Latin phrase *Quod erat demonstradum*. It translates as "Which was to be demonstrated."

Some Basic Laws

Logic operations obey certain rules, called *laws*. These laws are somewhat similar to the laws that govern the behavior of numbers in arithmetic, or variables in algebra. Following are some of the most basic laws of propositional logic.

PRECEDENCE

When reading or constructing logical statements, the operations within parentheses are always performed first. If there are multilayered combinations of sentences (called nesting of operations), then you should first use ordinary parentheses, then square brackets [], and then curly brackets {}. Alternatively, you can use groups of plain parentheses inside each other, but be sure you end up with the same number of left-hand parentheses and right-hand parentheses in the complete expression.

If there are no parentheses or brackets in an expression, instances of negation should be performed first. Then conjunctions should be done, then disjunctions, then implications, and finally logical equivalences.

As an example of how precedence works, consider the following compound sentence:

$$A \mathbin{\&} \neg B \vee C \Rightarrow D$$

Using parentheses and brackets to clarify this according to the rules of precedence, we can write it like this:

$$\{[A \mathbin{\&} (\neg B)] \vee C\} \Rightarrow D$$

Now consider a more complex compound sentence, which is so messy that we run out of parentheses and brackets if we use the "ordinary/square/curly" scheme:

$$A \mathbin{\&} \neg B \vee C \Rightarrow D \mathbin{\&} E \Leftrightarrow F \vee G$$

Using plain parentheses only, we can write it this way:

$$(((A \mathbin{\&} (\neg B)) \vee C) \Rightarrow (D \mathbin{\&} E)) \Leftrightarrow (F \vee G)$$

When we count up the number of left-hand parentheses and the number of right-hand parentheses, we see that there are six left-hand ones and six right-hand ones. (If the number weren't the same, we'd be in trouble!)

CONTRADICTION

A contradiction always results in a false truth value. This is one of the most interesting and useful laws in all of mathematics, and has been used to prove many important facts, as well as to construct satirical sentences. Symbolically, if X is any logical statement, we can write the rule like this:

$$(X \mathbin{\&} \neg X) \Rightarrow F$$

LAW OF DOUBLE NEGATION

The negation of a negation is equivalent to the original expression. That is, if X is any logical variable, then:

$$\neg(\neg X) \Leftrightarrow X$$

COMMUTATIVE LAWS

The conjunction of two variables always has the same value, regardless of the order in which the variables are expressed. If X and Y are logical variables, then X & Y is logically equivalent to Y & X:

$$X \& Y \Leftrightarrow Y \& X$$

The same property holds for logical disjunction:

$$X \vee Y \Leftrightarrow Y \vee X$$

These are called the *commutative law for conjunction* and the *commutative law for disjunction*, respectively. The variables can be commuted (interchanged in order) and it doesn't affect the truth value of the resulting sentence.

ASSOCIATIVE LAWS

When there are three variables combined by two conjunctions, it doesn't matter how the variables are grouped. Suppose you have a compound sentence that can be symbolized as follows:

$$X \& Y \& Z$$

where X, Y, and Z represent the truth values of three constituent sentences. Then we can consider X & Y as a single variable and combine it with Z, or we can consider Y & Z as a single variable and combine it with X, and the results are logically equivalent:

$$(X \& Y) \& Z \Leftrightarrow X \& (Y \& Z)$$

The same law holds for logical disjunction:

$$(X \vee Y) \vee Z \Leftrightarrow X \vee (Y \vee Z)$$

These are called the *associative law for conjunction* and the *associative law for disjunction*, respectively.

We must be careful when applying associative laws. All the operations in the compound sentence must be the same. If a compound sentence contains a conjunction and a disjunction, we cannot change the grouping and expect to get the same truth value in all possible cases. For example, the following two compound sentences are not, in general, logically equivalent:

$$(X \& Y) \vee Z$$
$$X \& (Y \vee Z)$$

LAW OF IMPLICATION REVERSAL

When one sentence implies another, you can't reverse the sense of the implication and still expect the result to be valid. It is not always true that if $X \Rightarrow Y$, then $Y \Rightarrow X$. It can be true in certain cases, such as when $X \Leftrightarrow Y$. But there are plenty of cases where it isn't true.

If you negate both sentences, then reversing the implication can be done, and the result is always valid. This is called the *law of implication reversal.* It is also known as the *law of the contrapositive*. Expressed symbolically, suppose we are given two logical variables X and Y. Then the following always holds:

$$(X \Rightarrow Y) \Leftrightarrow (\neg Y \Rightarrow \neg X)$$

PROBLEM 1-4

Use words to illustrate an example of the previous law in action, in a way that makes sense.

SOLUTION 1-4

Let V represent the sentence "Jane is a living vertebrate creature." Let B represent the sentence "Jane has a brain." Then $V \Rightarrow B$ reads, "If Jane is a living vertebrate creature, then Jane has a brain." Applying the law of implication reversal, we can also say with certainty that $\neg B \Rightarrow \neg V$. That translates to: "If Jane does not have a brain, then Jane is not a living vertebrate creature."

DeMORGAN'S LAWS

If the conjunction of two sentences is negated as a whole, the resulting compound sentence can be rewritten as the disjunction of the negations of the original two sentences. Expressed symbolically, if X and Y are two logical variables, then the following holds valid in all cases:

$$\neg(X \mathbin{\&} Y) \Leftrightarrow (\neg X \vee \neg Y)$$

This is called *DeMorgan's law for conjunction.*

A similar rule holds for disjunction. If a disjunction of two sentences is negated as a whole, the resulting compound sentence can be rewritten as the conjunction of the negations of the original two sentences. Symbolically:

$$\neg(X \vee Y) \Leftrightarrow (\neg X \ \& \ \neg Y)$$

This is called *DeMorgan's law for disjunction.*

You might now begin to appreciate the use of symbols to express complex statements in logic! The rigorous expression of *DeMorgan's laws* in verbal form is quite a mouthful, but it's easy to write these rules down as symbols.

DISTRIBUTIVE LAW

A specific relationship exists between conjunction and disjunction, known as the *distributive law*. It works somewhat like the distributive law that you learned in arithmetic classes—a certain way that multiplication behaves with respect to addition. Do you remember it? It states that if a and b are any two numbers, then

$$a(b + c) = ab + ac$$

Now think of logical conjunction as the analog of multiplication, and logical disjunction as the analog of addition. Then if X, Y, and Z are any three sentences, the following logical equivalence exists:

$$X \ \& \ (Y \vee Z) \Leftrightarrow (X \ \& \ Y) \vee (X \ \& \ Z)$$

This is called the *distributive law of conjunction with respect to disjunction.*

✓Truth Table Proofs

The laws of logic we've just stated were not merely dreamed up. They can be demonstrated to be true in general. Truth tables can be used for this purpose. If we claim that two compound sentences are logically equivalent, then we can show that their truth tables produce identical results. Also, if we can show that two compound sentences have truth tables that produce identical results, then we can be sure those two sentences are logically equivalent, as long as all possible combinations of truth values are accounted for.

The next few paragraphs show truth table proofs for the commutative laws, the associative laws, the law of implication reversal, DeMorgan's laws, and the distributive law. Some of these proofs seem trivial in their simplicity. When

Table 1-7. Truth table proof of the commutative law of conjunction. At A, statement of truth values for X & Y. At B, statement of truth values for Y & X. The outcomes are identical, demonstrating that they are logically equivalent.

A

X	Y	X & Y
F	F	F
F	T	F
T	F	F
T	T	T

B

X	Y	Y & X
F	F	F
F	T	F
T	F	F
T	T	T

some people see proofs like this, they ask, "Why bother with going through the motions, when these things are obvious?" The answer is this: In mathematics, something can appear to be obvious and then turn out to be false! In order to protect against mistaken conclusions, the pure mathematician adheres to a form of discipline called *rigor*. The following proofs are *rigorous*. They leave no room for doubt or dispute.

COMMUTATIVE LAW FOR CONJUNCTION

Tables 1-7A and 1-7B show that the following two general sentences are logically equivalent for any two variables X and Y:

$$X \,\&\, Y$$
$$Y \,\&\, X$$

COMMUTATIVE LAW FOR DISJUNCTION

Tables 1-8A and 1-8B show that the following two general sentences are logically equivalent for any two variables X and Y:

$$X \vee Y$$
$$Y \vee X$$

Table 1-8. Truth table proof of the commutative law of disjunction. At A, statement of truth values for X ∨ Y. At B, statement of truth values for Y ∨ X. The outcomes are identical, demonstrating that they are logically equivalent.

A

X	Y	X ∨ Y
F	F	F
F	T	T
T	F	T
T	T	T

B

X	Y	Y ∨ X
F	F	F
F	T	T
T	F	T
T	T	T

ASSOCIATIVE LAW FOR CONJUNCTION

Tables 1-9A and 1-9B show that the following two general sentences are logically equivalent for any three variables X, Y, and Z:

(X & Y) & Z
X & (Y & Z)

Table 1-9A. Derivation of truth values for (X & Y) & Z. Note that the last two columns of this proof make use of the commutative law for conjunction, which has already been proven.

A

X	Y	Z	X & Y	Z & (X & Y)	(X & Y) & Z
F	F	F	F	F	F
F	F	T	F	F	F
F	T	F	F	F	F
F	T	T	F	F	F
T	F	F	F	F	F
T	F	T	F	F	F
T	T	F	T	F	F
T	T	T	T	T	T

Table 1-9B. Derivation of truth values for X & (Y & Z). The far right-hand column of this table has values that are identical with those in the far right-hand column of Table 1-9A, demonstrating that the far right-hand expressions in the top rows are logically equivalent.

B

X	Y	Z	Y & Z	X & (Y & Z)
F	F	F	F	F
F	F	T	F	F
F	T	F	F	F
F	T	T	T	F
T	F	F	F	F
T	F	T	F	F
T	T	F	F	F
T	T	T	T	T

Note that in Table 1-9A, the last two columns make use of the commutative law for conjunction, which has already been proven. Once proven, a statement is called a *theorem*, and it can be used in future proofs.

ASSOCIATIVE LAW FOR DISJUNCTION

Tables 1-10A and 1-10B show that the following two general sentences are logically equivalent for any three variables X, Y, and Z:

$$(X \vee Y) \vee Z$$
$$X \vee (Y \vee Z)$$

In Table 1-10A, we take advantage of the commutative law for disjunction, which has already been proved, in the last two columns.

Table 1-10A. Derivation of truth values for (X ∨ Y) ∨ Z. Note that the last two columns of this proof make use of the commutative law for disjunction, which has already been proven.

A

X	Y	Z	X ∨ Y	Z ∨ (X ∨ Y)	(X ∨ Y) ∨ Z
F	F	F	F	F	F
F	F	T	F	T	T
F	T	F	T	T	T
F	T	T	T	T	T
T	F	F	T	T	T
T	F	T	T	T	T
T	T	F	T	T	T
T	T	T	T	T	T

Table 1-10B. Derivation of truth values for X ∨ (Y ∨ Z). The far right-hand column of this table has values that are identical with those in the far right-hand column of Table 1-10A, demonstrating that the far right-hand expressions in the top rows are logically equivalent.

B

X	Y	Z	Y ∨ Z	X ∨ (Y ∨ Z)
F	F	F	F	F
F	F	T	T	T
F	T	F	T	T
F	T	T	T	T
T	F	F	F	T
T	F	T	T	T
T	T	F	T	T
T	T	T	T	T

LAW OF IMPLICATION REVERSAL

Tables 1-11A and 1-11B show that the following two general sentences are logically equivalent for any two variables X and Y:

$$X \Rightarrow Y$$
$$\neg Y \Rightarrow \neg X$$

Table 1-11. Truth table proof of the law of implication reversal. At A, statement of truth values for $X \Rightarrow Y$. At B, derivation of truth values for $\neg Y \Rightarrow \neg X$. The outcomes are identical, demonstrating that they are logically equivalent.

A

X	Y	X ⇒ Y
F	F	T
F	T	T
T	F	F
T	T	T

B

X	Y	¬Y	¬X	¬Y ⇒ ¬X
F	F	T	T	T
F	T	F	T	T
T	F	T	F	F
T	T	F	F	T

DeMORGAN'S LAW FOR CONJUNCTION

Tables 1-12A and 1-12B show that the following two general sentences are logically equivalent for any two variables X and Y:

$$\neg(X \;\&\; Y)$$
$$\neg X \vee \neg Y$$

Table 1-12. Truth table proof of DeMorgan's law for conjunction. At A, statement of truth values for ¬(X & Y). At B, derivation of truth values for ¬X ∨ ¬Y. The outcomes are identical, demonstrating that they are logically equivalent.

A

X	Y	X & Y	¬(X & Y)
F	F	F	T
F	T	F	T
T	F	F	T
T	T	T	F

B

X	Y	¬X	¬Y	¬X ∨ ¬Y
F	F	T	T	T
F	T	T	F	T
T	F	F	T	T
T	T	F	F	F

DeMORGAN'S LAW FOR DISJUNCTION

Tables 1-13A and 1-13B show that the following two general sentences are logically equivalent for any two variables X and Y:

$$\neg(X \vee Y)$$
$$\neg X \; \& \; \neg Y$$

Table 1-13. Truth table proof of DeMorgan's law for disjunction. At A, statement of truth values for $\neg(X \vee Y)$. At B, derivation of truth values for $\neg X$ & $\neg Y$. The outcomes are identical, demonstrating that they are logically equivalent.

A

X	Y	X ∨ Y	¬(X ∨ Y)
F	F	F	T
F	T	T	F
T	F	T	F
T	T	T	F

B

X	Y	¬X	¬Y	¬X & ¬Y
F	F	T	T	T
F	T	T	F	F
T	F	F	T	F
T	T	F	F	F

DISTRIBUTIVE LAW

Tables 1-14A and 1-14B show that the following two general sentences are logically equivalent for any three variables X, Y, and Z:

$$X \; \& \; (Y \vee Z)$$
$$(X \; \& \; Y) \vee (X \; \& \; Z)$$

Table 1-14A. Derivation of truth values for X & (Y ∨ Z).

A

X	Y	Z	Y ∨ Z	X & (Y ∨ Z)
F	F	F	F	F
F	F	T	T	F
F	T	F	T	F
F	T	T	T	F
T	F	F	F	F
T	F	T	T	T
T	T	F	T	T
T	T	T	T	T

Table 1-14B. Derivation of truth values for (X & Y) ∨ (X & Z). The far right-hand column of this table has values that are identical with those in the far right-hand column of Table 1-14A, demonstrating that the far right-hand expressions in the top rows are logically equivalent.

B

X	Y	Z	X & Y	X & Z	(X & Y) ∨ (X & Z)
F	F	F	F	F	F
F	F	T	F	F	F
F	T	F	F	F	F
F	T	T	F	F	F
T	F	F	F	F	F
T	F	T	F	T	T
T	T	F	T	F	T
T	T	T	T	T	T

PROBLEM 1-5

Using truth tables, prove the following logical proposition:

$$[(X \,\&\, Y) \Rightarrow Z] \Leftrightarrow [\neg Z \Rightarrow (\neg X \vee \neg Y)]$$

SOLUTION 1-5

Tables 1-15A and 1-15B show that the following two general sentences are logically equivalent for any three variables X, Y, and Z:

$$(X \,\&\, Y) \Rightarrow Z$$
$$\neg Z \Rightarrow (\neg X \vee \neg Y)$$

PROBLEM 1-6

Use rules that we have presented in this chapter, rather than truth-table comparison, to prove the proposition stated in Problem 1-5.

SOLUTION 1-6

First, consider DeMorgan's law for conjunction. This states that the following two sentences are logically equivalent for any X and Y:

$$\neg(X \,\&\, Y)$$
$$\neg X \vee \neg Y$$

Table 1-15A. Derivation of truth values for (X & Y) ⇒ Z.

A

X	Y	Z	X & Y	(X & Y) ⇒ Z
F	F	F	F	T
F	F	T	F	T
F	T	F	F	T
F	T	T	F	T
T	F	F	F	T
T	F	T	F	T
T	T	F	T	F
T	T	T	T	T

Table 1-15B. Derivation of truth values for $\neg Z \Rightarrow (\neg X \vee \neg Y)$. The far right-hand column of this table has values that are identical with those in the far right-hand column of Table 1-15A, demonstrating that the far right-hand expressions in the top rows are logically equivalent.

B

X	Y	Z	$\neg X$	$\neg Y$	$\neg Z$	$\neg X \vee \neg Y$	$\neg Z \Rightarrow (\neg X \vee \neg Y)$
F	F	F	T	T	T	T	T
F	F	T	T	T	F	T	T
F	T	F	T	F	T	T	T
F	T	T	T	F	F	T	T
T	F	F	F	T	T	T	T
T	F	T	F	T	F	T	T
T	T	F	F	F	T	F	F
T	T	T	F	F	F	F	T

This means that the two expressions are directly interchangeable. Whenever we encounter either of these in any logical sentence, we can "pull it out" and "plug in" the other one. Let's take advantage of this fact on the right-hand side of the second expression in Problem 1-5, changing it to the following:

$$\neg Z \Rightarrow \neg(X \,\&\, Y)$$

According to the law of implication reversal, this is logically equivalent to:

$$\neg[\neg(X \,\&\, Y)] \Rightarrow \neg(\neg Z)$$

Using the law of double negation on both sides of this expression, we see that this is logically equivalent to:

$$(X \,\&\, Y) \Rightarrow Z$$

This is precisely the first expression in Problem 1-5. This shows that the first and second expressions in Problem 1-5 are logically equivalent.

Quiz

This is an "open book" quiz. You may refer to the text in this chapter. A good score is eight correct. Answers are in the back of the book.

1. The conjunction of three sentences is false
   (a) if and only if all three sentences are false.
   (b) if and only if at least one of the sentences is false.
   (c) if and only if at least two of the sentences are false.
   (d) under no circumstances, because a conjunction can't be defined for more than two sentences.

2. The disjunction of three sentences is false
   (a) if and only if all three sentences are false.
   (b) if and only if at least one of the sentences is false.
   (c) if and only if at least two of the sentences are false.
   (d) under no circumstances, because a disjunction can't be defined for more than two sentences.

3. In a logical implication, the double-shafted arrow pointing to the right can be replaced by the word or words
   (a) "and."
   (b) "if."
   (c) "if and only if."
   (d) "implies."

4. How many possible combinations of truth values are there for a set of three sentences, each of which can attain either the value T or the value F?
   (a) 2
   (b) 4
   (c) 8
   (d) 16

5. Suppose you observe, "It is not sunny today, and it's not warm." Your friend says, "The statement that it's sunny or warm today is false." These two sentences are logically equivalent, and this constitutes a verbal example of
   (a) one of DeMorgan's laws.
   (b) the law of double negation.
   (c) the commutative law for conjunction.
   (d) the law of implication reversal.

6. Imagine that I claim a certain general statement is a rule of logic. You demonstrate that my supposed rule has at least one exception. This shows that
   (a) it is not a law of logic.
   (b) it violates the commutative law.
   (c) it violates the law of implication reversal.
   (d) it demonstrates that a disjunction implies logical falsity.

7. Look at Table 1-16. What, if anything, is wrong with this truth table?
   (a) Not all possible combinations of truth values are shown for X, Y, and Z.
   (b) The entries in the far right-hand column are incorrect.
   (c) It is impossible to have a logical operation such as (X ∨ Y) & Z.
   (d) Nothing is wrong with Table 1-16.

8. What, if anything, can be done to make Table 1-16 show a valid derivation?
   (a) Nothing needs to be done. It is correct as it is.
   (b) In the top row, far-right column header, change the ampersand (&) to a double-shafted arrow pointing to the right (⇒).
   (c) In the far-left column, change every T to an F, and change every F to a T.
   (d) In the first three columns, change every T to an F, and change every F to a T.

Table 1-16. Truth table for Quiz Questions 7 and 8.

A

X	Y	Z	X ∨ Y	(X ∨ Y) & Z
F	F	F	F	T
F	F	T	F	T
F	T	F	T	F
F	T	T	T	T
T	F	F	T	F
T	F	T	T	T
T	T	F	T	F
T	T	T	T	T

9. A rule or law that has been proven
   (a) can't be used to prove future theorems, because all theorems must be proven directly from an original set of rules.
   (b) can be used to prove future theorems, as long as truth tables are avoided.
   (c) can be used to prove future theorems, but only by means of truth tables.
   (d) can be used to prove future theorems.

10. Imagine that someone says to you, "If I am a human and I am not a human, then the moon is made of Swiss cheese." (Forget for a moment that this person has obviously lost contact with the real world.) This is a verbal illustration of the fact that
    (a) implication can't be reversed.
    (b) DeMorgan's laws don't always hold true.
    (c) conjunction is not commutative.
    (d) a contradiction implies logical falsity.

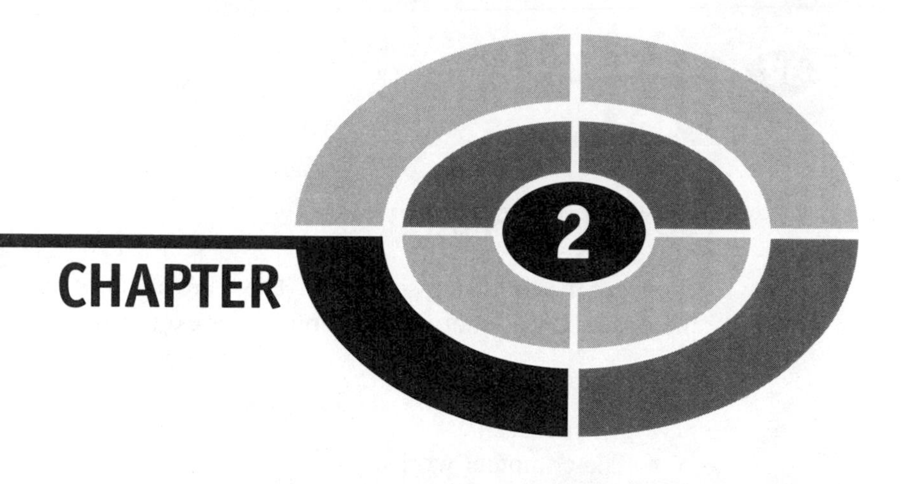

How Sentences Are Put Together

Plenty of things can be proved without dissecting sentences into smaller parts. But sometimes, it's necessary to break sentences down. That's what *predicate logic*, also called *predicate calculus*, is all about. In this chapter, we'll see how sentences should (and should not) be constructed.

Sentence Structure

When a sentence (proposition) takes the form of a declaration, that sentence can be split into a *subject* and a *predicate*. The subject is a *noun*, or "naming word." It is the center of attention in the sentence. The predicate gives information about the subject. This information can be "passive," such as a description of the subject's color or shape, or "active," such as an expression of what the subject does or where it goes.

SUBJECT/VERB (SV)

Consider the following sentences. These are about as basic as sentences get (with the exception of one-word commands or exclamations):

- Jack walks.
- Jill sneezes.
- The computer works.
- You shop.

Each of these sentences contains a noun (the subject) followed by a *verb*—an "action word"—and that is all. The action is not directed at anything in particular, nor does it happen in any special way. "Jack walks." Where? To school, in his home, or through the woods? Does he walk fast or slowly? We don't know. "Jill sneezes." Does she sneeze at you, or at the wall, or into a handkerchief? Does she sneeze loudly or quietly? No clue. "The computer works." How well? How fast? With which programs? We are not told. "You shop." For what? Where? For how long? Not specified. The sentences are vague, but they are nevertheless well-formed propositions. They are called *subject/verb* (SV) sentences.

SUBJECT/VERB/OBJECT (SVO)

Consider the following sentences:

- Jack walks to school.
- Jill kicks the ball.
- I mow the lawn.
- You trim a tree.

Each of these sentences contains a noun (the subject) followed by a verb, and then there is another noun that is influenced or acted upon by the subject and verb. In all four of these sentences, the subjects are people: Jack, Jill, I, and you. (Subjects don't have to be people, or even animate things, however.) The objects in the previous examples are inanimate: school, ball, lawn, and tree. (This, too, is a coincidence. Objects aren't always inanimate.) Each sentence also contains a verb: walk, kick, mow, and trim.

Let S represent a subject, V a verb, and O an object. We can diagram each of the above sentences as shown in Fig. 2-1. The subject, by means of the verb, per-

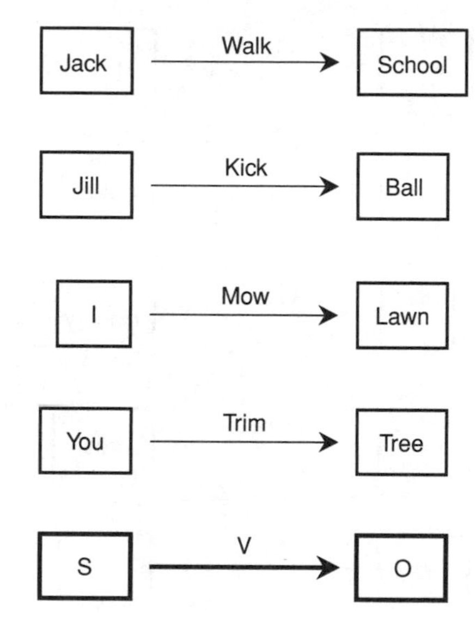

Fig. 2-1. Examples of subject/verb/object (SVO) sentences.

forms some action on the object. This type of statement can be called a *subject/verb/object* (SVO) sentence.

SUBJECT/LINKING VERB/COMPLEMENT

Now look at the following sentences:

- Jack is a boy.
- Jill has a cold.
- I was hungry.
- You will get tired.

Each of these sentences contains a noun (the subject) at the beginning. Then there's a word at the end that tells us some detail about the subject; it complements the subject. The subject and the *complement* are linked by a word in the middle, which we'll call a *link* or *linking verb*.

If we let S represent a subject, LV a linking verb, and C a complement, then we can diagram each of the previous sentences as shown in Fig. 2-2. These are examples of *subject/linking verb/complement* (SLVC) sentences.

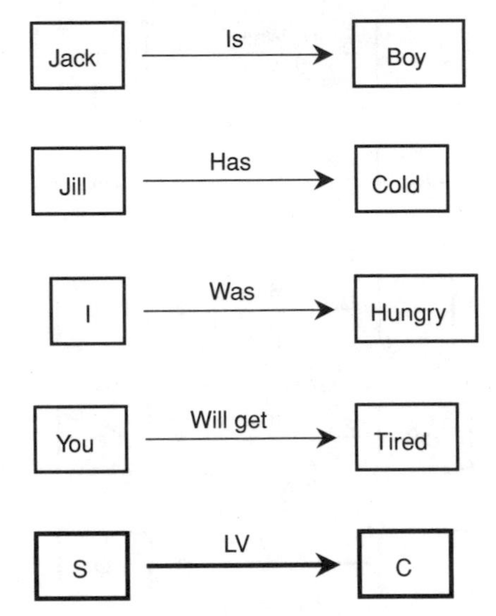

Fig. 2-2. Examples of subject/linking verb/complement (SLVC) sentences.

WHAT'S THE PREDICATE?

In an SVO sentence, the predicate consists of the verb and the object. In an SLVC sentence, the predicate consists of the linking verb and the complement.

The eight sentences in the previous two sections can be broken into subjects and predicates, as shown in Table 2-1. The subjects and predicates are assigned symbols. Predicates are given non-italicized uppercase letters of the alphabet, and subjects are given italicized lowercase letters. These assignments seem arbitrary, but that's all right as long as we agree on them. (Ideally we'd use the first letter of each noun, but some of these coincide here, and that could cause confusion among the sentences.)

Using the symbols in the table, we can denote each of the sentences by writing the symbol for the predicate first, followed by the symbol for the subject. The first four sentences, which are SVO type, are thus denoted W*a*, K*i*, M*q*, and T*u*. The second four sentences, which have the same subjects but are of the SLVC form, are denoted B*a*, C*i*, H*q*, and R*u*. In all eight of the sentences, the symbols for the subjects are called *logical constants* (or simply *constants*) because they denote specific, identifiable subjects.

Table 2-1. Sentences broken into subjects and predicates, along with symbols. Symbols are arbitrary to prevent coincidences that could cause confusion among the sentences.

Subject	Subject Symbol	Predicate	Predicate Symbol
Jack	*a*	walks to school	W
Jill	*i*	kicks the ball	K
I	*q*	mow the lawn	M
You	*u*	trim a tree	T
Jack	*a*	is a boy	B
Jill	*i*	has a cold	C
I	*q*	was hungry	H
You	*u*	will get tired	R

VARIABLES

Now imagine that, instead of specific nouns as subjects, we use unspecified ones. You ask, "What is an unspecified noun?" The answer is, "Anything, as long as we don't say exactly what." In predicate logic, a non-specific noun is called a *logical variable* (or simply a *variable*), and is represented by a lowercase letter in italics, usually from near the end of the alphabet. A favorite is *x*. The letters *w*, *y*, and *z* are also popular for symbolizing variables. If necessary, subscripts can be used if a sentence has a lot of variables: for example, x_1, x_2, x_3, x_4, x_5, and so on.

Examine the generalized sentences shown in Table 2-2. The constants *a*, *i*, *q*, and *u* have been replaced by the variable *x*. But the structures of the sentences in Table 2-2 are identical to their counterparts on corresponding lines of Table 2-1. There is no need to use different letters for the subjects in Table 2-2, because the variable is non-specific by nature. There can't be any confusion among things or people when we don't say exactly what or who they are!

When we replace the constants with the variable *x*, we have eight sentences that can be denoted as follows: W*x*, K*x*, M*x*, T*x*, B*x*, C*x*, H*x*, and R*x*. In every case, the predicate, symbolized by a non-italicized uppercase letter, tells us

Table 2-2. Sentences broken into subjects and predicates, along with symbols. The subject in each case is "someone" and is represented by the variable x.

Subject	Subject Symbol	Predicate	Predicate Symbol
Someone	x	walks to school	W
Someone	x	kicks the ball	K
Someone	x	mows the lawn	M
Someone	x	trims a tree	T
Someone	x	is a boy	B
Someone	x	has a cold	C
Someone	x	was hungry	H
Someone	x	will get tired	R

something about the variable, symbolized by x. The first four predicates are "doing" expressions, and the second four are "being" expressions. In a general sense, we can say "x does or goes to such-and-such" for the first four sentences, and "x has such-and-such a characteristic" for the second four sentences.

PROBLEM 2-1

What types of sentences are the following? Identify their parts.

- The atmosphere has layers.
- The tornado destroyed the barn.
- I bought a computer.
- My computer is defective.

SOLUTION 2-1

The first sentence is of the SLVC type. The subject is "atmosphere," the linking verb is "has," and the complement is "layers."

The second sentence is SVO. The subject is "tornado," the verb is "destroyed," and the object is "barn."

The third sentence is SVO. The subject is "I," the verb is "bought," and the object is "computer."

The fourth sentence is SLVC. The subject is "my computer," the linking verb is "is," and the complement is "defective."

PROBLEM 2-2
Evaluate the fourth sentence in the previous problem in another way, and write down a symbolic expression for it.

SOLUTION 2-2
This sentence can be considered the negation of the SLVC sentence "My computer is perfect." If we symbolize "my computer" by c and "perfect" by P, then "My computer is perfect" can be denoted $\mathrm{P}c$, and "It is not true that my computer is perfect" can be denoted $\neg(\mathrm{P}c)$.

PROBLEM 2-3
Identify the predicate in each of the sentences stated in Problem 2-1.

SOLUTION 2-3
The predicates are "has layers," "destroyed the barn," "bought a computer," and "is defective," respectively. If we consider the fourth sentence as the equivalent of "It is not true that my computer is perfect," then the predicate of the negated sentence becomes "is perfect."

Quantifiers

The foregoing sentences are much simpler than most of the things people say. Let's go to the next level of complexity.

UNIVERSAL QUANTIFIER

Look at the following sentences. The first two are SVO, and the second two are SLVC. But all four of these sentences have something in common.

- Every boy walks to school.
- Every football gets kicked.
- All swimmers are hungry.
- All teachers are geniuses.

The common feature of the above sentences is the fact that they are blanket statements. They speak about things as being universally true (or false). They can be reworded like this:

- For any thing, if that thing is a boy, then that thing walks to school.
- For any thing, if that thing is a football, then that thing gets kicked.
- For any thing, if that thing is a swimmer, then that thing is hungry.
- For any thing, if that thing is a teacher, then that thing is a genius.

Each subject has become a variable, represented by the generic word or phrase "thing" or "that thing." Let's replace the word "thing" and the phrase "that thing" in each of the above statements by the symbol x. Here's what we get:

- For any x, if x is a boy, then x walks to school.
- For any x, if x is a football, then x gets kicked.
- For any x, if x is a swimmer, then x is hungry.
- For any x, if x is a teacher, then x is a genius.

There is a symbol in predicate logic that stands for the words "for all," "for every," or "for any." That symbol is $\forall$. It looks like an upside-down uppercase letter A, and is called the *universal quantifier* because it indicates that something is universally true about a variable. The variable to which the quantifier applies is written right after the symbol.

Now let's symbolize the phrases in the above sentences according to Table 2-3. If we write out the sentences symbolically, using the $\Rightarrow$ symbol from propositional logic to indicate implication, we get these:

$$(\forall x)\ \mathrm{B}x \Rightarrow \mathrm{W}x$$
$$(\forall x)\ \mathrm{F}x \Rightarrow \mathrm{K}x$$
$$(\forall x)\ \mathrm{S}x \Rightarrow \mathrm{H}x$$
$$(\forall x)\ \mathrm{T}x \Rightarrow \mathrm{G}x$$

The quantifier is placed in parentheses to separate it from the sentence that follows. There are other ways a universal quantifier can be set apart from the rest of the sentence: using a colon after $\forall x$, using a vertical line after $\forall x$, and placing a portion of the sentence after the quantifier in parentheses while not using parentheses around the quantifier. Therefore, for example, we can write any of the following to represent the first of the above symbolized sentences:

$$\forall x\text{: } \mathrm{B}x \Rightarrow \mathrm{W}x$$
$$\forall x \mid \mathrm{B}x \Rightarrow \mathrm{W}x$$
$$\forall x\ (\mathrm{B}x \Rightarrow \mathrm{W}x)$$

Table 2-3. Predicate symbols used to denote some sentences containing quantifiers of a variable.

Predicate	Predicate Symbol
is a boy	B
walks to school	W
is a football	F
gets kicked	K
is a swimmer	S
is hungry	H
is a teacher	T
is a genius	G

These two notations are commonly used in mathematics papers and texts. The parentheses are more often used when writing about predicate logic.

EXISTENTIAL QUANTIFIER

Now examine the following sentences. They apply to the same subjects, verbs, linking verbs, objects, and complements as the sentences in the preceding paragraphs. But there is an important difference! These sentences are not blanket statements. In fact, there is a definite suggestion that there are exceptions to the rules:

- Some boys walk to school.
- Some footballs get kicked.
- Some swimmers are hungry.
- Some teachers are geniuses.

These sentences can be reworded to get the following:

- There exists a thing, such that if that thing is a boy, then that thing walks to school.
- There exists a thing, such that if that thing is a football, then that thing gets kicked.
- There exists a thing, such that if that thing is a swimmer, then that thing is hungry.
- There exists a thing, such that if that thing is a teacher, then that thing is a genius.

Again, each subject has become a variable, represented by the generic word or phrase "thing" or "that thing." Let's replace the word "thing" and the phrase "that thing" in each of the above statements by the symbol x. Here's what we get:

- There exists an x, such that if x is a boy, then x walks to school.
- There exists an x, such that if x is a football, then x gets kicked.
- There exists an x, such that if x is a swimmer, then x is hungry.
- There exists an x, such that if x is a teacher, then x is a genius.

Logicians use a symbol to stand for the words "there exists," "there is," "for some," or "for at least one." That symbol is ∃, a backwards uppercase letter E. It is called the *existential quantifier*. It indicates that something can be, or sometimes is, true about a variable. The variable to which the quantifier applies is, as with the universal quantifier, written right after the symbol.

Refer to Table 2-3 and symbolize the sentence parts. If we write out the foregoing existential-quantifier sentences symbolically, using the ⇒ symbol from propositional logic to indicate implication, we get:

$$(\exists x)\ \mathrm{B}x \Rightarrow \mathrm{W}x$$
$$(\exists x)\ \mathrm{F}x \Rightarrow \mathrm{K}x$$
$$(\exists x)\ \mathrm{S}x \Rightarrow \mathrm{H}x$$
$$(\exists x)\ \mathrm{T}x \Rightarrow \mathrm{G}x$$

The quantifier is, again, placed in parentheses to separate it from the sentence that follows. As with the universal quantifier, we can have these alternative notations for the first of the above sentences:

$$\exists x\text{: } \mathrm{B}x \Rightarrow \mathrm{W}x$$
$$\exists x \mid \mathrm{B}x \Rightarrow \mathrm{W}x$$
$$\exists x\ (\mathrm{B}x \Rightarrow \mathrm{W}x)$$

If there is any doubt about which portion of the sentence after a quantifier is "covered" by that quantifier, then parentheses or higher-order brackets should be placed around only that portion of the sentence affected. For example, suppose you write this:

$$(\forall x)\ Px\ \&\ Qx\ \&\ Rx$$

This could be confused with either of the following:

$$(\forall x)\ Px\ \&\ (Qx\ \&\ Rx)$$
$$(\forall x)\ (Px\ \&\ Qx\ \&\ Rx)$$

If there aren't any parentheses around an expression following a quantifier, you should interpret this to mean that the entire expression is to be considered as a whole. Thus, the second of the above interpretations is the correct one. But here's a good rule you can follow when working with logical formulas: "When in doubt, it's better to have too many sets of parentheses than not enough." That is, it's better to clarify oneself excessively than insufficiently!

MULTIPLE QUANTIFIERS

It's possible to have sentences in which there is more than one quantifier, each one applying to a different variable. For example:

$$(\forall x)(\exists y)(\exists z)\ (Px\ \&\ Qy\ \&\ Rz)$$

This is read as follows: "For all x, there exists a y and there exists a z such that Px and Qy and Rz."

If all the quantifiers are of the same type (either universal or existential) in a multiple-quantifier expression, then the quantifier can be listed once, and all the applicable variables can be listed following it. For example:

$$(\forall x, y, z)\ (Px\ \&\ Qy\ \&\ Rz)$$

This is read as follows: "For all x, for all y, and for all z, Px and Qy and Rz."

PROBLEM 2-4

Symbolize the following sentences:

- Caesar is a human being.
- All human beings will die.
- Caesar will die.

SOLUTION 2-4

Let H represent the predicate "is a human being." Let D represent the predicate "will die." Let c represent the subject "Caesar," which is a constant. Let x represent a logical variable. Then the above sentences can be symbolized:

$$Hc$$
$$(\forall x)\ Hx \Rightarrow Dx$$
$$Dc$$

PROBLEM 2-5

Symbolize the following logical argument. "Caesar is a human being. All human beings will die. Therefore, Caesar will die."

SOLUTION 2-5

We already have the symbolic representations of the three sentences contained in the argument. This argument states that if the first two sentences are both true, then the last one is true. So we can write the argument like this:

$$Hc \ \&\ [(\forall x)\ Hx \Rightarrow Dx] \Rightarrow Dc$$

Remember the rules for precedence outlined in the last chapter. Tasks within parentheses or brackets are performed first. Then conjunction is performed, and then implication. If you want to use extra brackets to avoid any possibility of confusion, you can write the argument this way:

$$\{Hc \ \&\ [(\forall x)\ Hx \Rightarrow Dx]\} \Rightarrow Dc$$

Well-Formed Formulas

In propositional logic, every sentence is written as a single symbol. Such a symbol can't be put down with incorrect structure, because it's a self-contained whole. But in predicate logic, sentences are broken down into parts. This means they must have a certain syntax, just as the sentences you utter or write in everyday life should obey certain rules of grammar (or *would*, in an English teacher's paradise).

WHY BOTHER WITH SYNTAX?

In every generation, new grammar rules evolve. Sentences that would have given a language purist nightmares 50 years ago are commonplace today. A few decades in the future, some of the sentences we think are fashionable now will sound archaic, foreign, or stupid. Nonstandard syntax can "come out goofy." That's not necessarily a major problem in casual speech. But in logic, sloppy syntax cannot be allowed, because it produces meaningless, ambiguous, or inaccurate statements.

Problems with syntax have caused serious misunderstandings between people in cultures where the sentences are not put together in the same ways. Problems with syntax can also cause intergenerational conflicts. We don't want that sort of thing to happen in logic. We can't afford to allow any room for confusion when we want to prove something!

WHAT IS A WFF?

A properly constructed sentence in predicate logic, translated into symbols according to certain rules, is called a *well-formed formula*. This term is often abbreviated *wff* (pronounced "wiff" or "woof").

Let's use boldface uppercase letters from the latter part of the alphabet (such as **X**) to represent unspecified subject/predicate sentences. There are certain basic rules for constructing such sentences. Here they are:

- All sentences in propositional logic are wffs.
- If A is a predicate and k is a constant or variable, then Ak is a wff. In other words, any predicate can be put together with any subject, and the result is a wff.
- If A is a predicate and k_1, k_2, k_3,..., and k_n are constants or variables, then A$k_1k_2k_3\ldots k_n$, representing the conjunction Ak_1 & Ak_2 & Ak_3 &...& Ak_n, is a wff. In other words, a predicate can apply to more than one subject.
- If A is a predicate, k_1, k_2, k_3,..., and k_n are constants or variables, and we are given a wff of the form A$k_1k_2k_3\ldots k_n$, then Ak_1, Ak_2, Ak_3,... and Ak_n are all wffs. In other words, if a wff contains a predicate and multiple subjects, then that predicate can be put together with any one of the subjects, and the result is a wff.
- If **X** is any wff containing the variables x_1, x_2, x_3,... x_n that do not have quantifiers, and if we let a quantifier (either universal or existential) be represented by the "wild card" symbol #, then $(\#x_1)(\#x_2)(\#x_3)\ldots(\#x_n)$ **X** is a wff.
- Sentential negation, conjunction, disjunction, implication, and logical equivalence can all be used with or between predicate wffs, just as they can

be used with or between simple propositions, and the result is always a wff.
- Any formula that does not conform to these rules is not a wff.

Here's an important thing to remember: A statement does not have to be true in order to be a wff! Statements whose truth is not known, or that are obviously false, can nevertheless be perfect wffs.

If the above syntax rules seem complicated, read them through a few times. All they are meant to do is tell us how to put sentences together so they make logical (if not always common) sense.

MULTIPLE CONSTANTS

Thus far, we've constructed wffs in which there is only one constant or one variable. However, in the above rules, there is mention of multiple constants. In most such cases, there are two constants: the subject and the object in an SVO sentence. The option of symbolizing sentences with multiple constants lets us express things in more detail than would be possible if multiple constants were not allowed.

Consider the following sentences:

- Jill walks to school.
- Bob kicks the football.
- That runner eats pork.
- My teacher understands Einstein.

Let's symbolize the nouns and verbs as indicated in Table 2-4. We can then write the sentences like this:

$$Wjs$$
$$Kbf$$
$$Erp$$
$$Ute$$

We list the verb first, then the subject, and then the object. The order in which the constants appear is important.

Suppose we reverse the order of the constants in each of the above sentences? Then we get the following:

$$Wsj$$
$$Kfb$$
$$Epr$$
$$Uet$$

Table 2-4. Nouns and verbs used to denote some sentences containing two constants.

Nouns	**Symbols**
Jill	j
Bob	b
that runner	r
my teacher	t
school	s
the football	f
pork	p
Einstein	e
Verbs	**Symbols**
walks to	W
kicks	K
eats	E
understands	U

Assuming we keep the symbol assignments shown in Table 2-4, these symbolic representations translate this way:

- The school walks to Jill.
- The football kicks Bob.
- Pork eats that runner.
- Einstein understands my teacher.

These are legitimate wffs, even though they come out strange when expanded into words. Nothing in the syntax rules forbids a wff to be ridiculous when translated into everyday language. (Imprecision or ambiguity is intolerable, but silliness is all right.)

Multiple variables are allowable, too. If we don't want to specify the constants in the preceding four wffs, we might use variables x and y instead, getting these:

W*xy*
K*xy*
E*xy*
U*xy*

These wffs translate like this, respectively:

- *x* walks to *y*.
- *x* kicks *y*.
- *x* eats *y*.
- *x* understands *y*.

PROBLEM 2-6
Which of the following are wffs? Which are not? Variables are symbolized as *x* and *y*. Constants are symbolized as *a* and *b*. Predicates are symbolized as R and S.

*x*R*ab*
S*abx*
RS*a*
R*xy*S*b*
*a*S

SOLUTION 2-6
Only the second expression, S*abx*, conforms to the syntax rules for predicate wffs. Therefore, it alone is a wff.

PROBLEM 2-7
Write out the second sentence above (a legitimate wff) using the words indicated in Table 2-5 in place of the predicate, constants, and variable.

SOLUTION 2-7
Here it is! Note that the first symbol in a wff always represents a verb or predicate, and should be treated as such.

- Adam stands in front of Betsy and a person from France.

PROBLEM 2-8
In order to illustrate, in words, what can happen when predicate formulas do not conform to the rules for wffs, write out the faulty formulas from Problem 2-6, using the words indicated in Table 2-5 in place of

Table 2-5. Table for Problems 2-7 and 2-8.

Symbol	Word of Phrase
R	runs towards
S	stands in front of
a	Adam
b	Betsy
x	a person from France
y	a person from England

the predicates, constants, and variables. Note that the first symbol in a wff always represents a verb or predicate, and should be treated as such.

SOLUTION 2-8

The following represent good attempts, at least, to translate the four faulty wffs shown previously.

- Runs toward a person from France Adam and Betsy.
- Stands in front of runs toward Adam.
- A person from France runs toward a person from England, stands in front of, and Betsy.
- Stands in front of Adam-ifies.

There are undoubtedly other ways to expand the faulty formulas (which we might call poorly-formed formulas, or pffs). But from these, you should get the idea that sentence construction is important. Even the most fervent rebels against the English language will wrinkle their noses at mutilated sentences like these for decades to come!

Venn Diagrams

Quantifiers can be applied to sets of subjects or objects, and these relationships can be illustrated as *Venn diagrams*. Let's look at some examples. Imagine two types of things, known as *widgets* and *doodads*. Imagine two sets, one consisting of all the widgets in the world, and the other consisting of all the doodads in the world. Suppose that both sets contain lots of items.

SOME (MAYBE ALL) WIDGETS ARE DOODADS

Let the predicate W represent "is a widget," and the predicate D represent "is a doodad." Consider this sentence:

$$(\exists x)\ (Wx\ \&\ Dx)$$

This translates as, "There exists at least one x, such that x is a widget and x is a doodad." It can also be translated as "Some (maybe all) widgets are doodads." There are four ways this can occur, as shown in Fig. 2-3. The set of widgets is represented by the solid rectangle and its interior. The set of doodads is represented by the dashed rectangle and its interior. Individual widgets and doodads can be represented by points inside the respective rectangles.

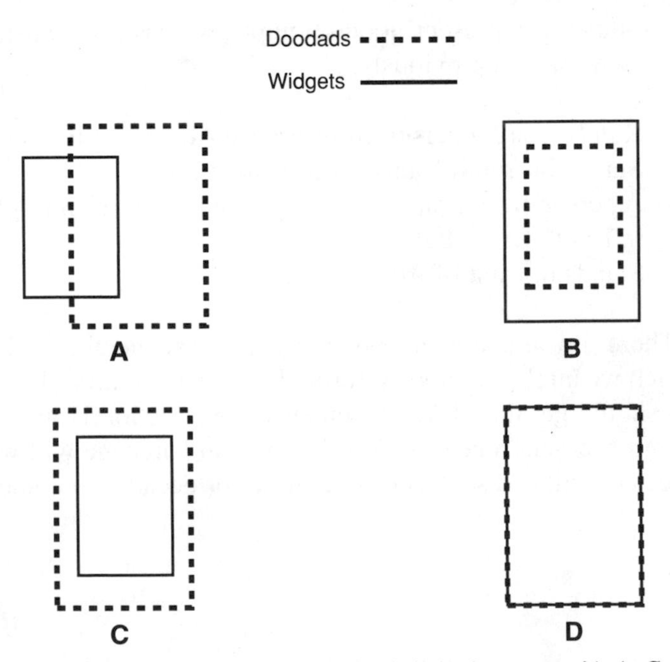

Fig. 2-3. At A and B, some widgets are doodads (and some aren't). At C, some widgets are doodads (in fact, they all are). At D, some widgets are doodads (in fact, the set of widgets and the set of doodads coincide).

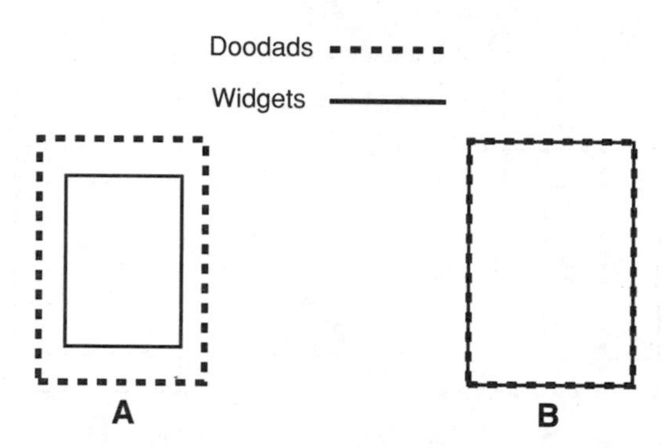

Fig. 2-4. At A, all widgets are doodads (but some doodads aren't widgets). At B, all widgets are doodads (and all doodads are widgets, too).

ALL WIDGETS ARE DOODADS

Now suppose that all widgets are doodads. The technical way to state this is, "For every x, if x is a widget, then x is a doodad." This is written as follows in symbolic language:

$$(\forall x)\ (\mathrm{W}x \Rightarrow \mathrm{D}x)$$

This can happen in two different ways, as shown in Fig. 2-4.

SOME (BUT NOT ALL) WIDGETS ARE DOODADS

Now let's go back to the situation shown in Fig. 2-3, but place a constraint on it. We now say, "There exists at least one x, such that x is a widget and x is a doodad. But it is not true that for all x, x is a widget and x is a doodad." More simply, we would say, "Some (but not all) widgets are doodads." This can be symbolized as follows:

$$[(\exists x)\ (\mathrm{W}x\ \&\ \mathrm{D}x)]\ \&\ \neg[(\forall x)\ (\mathrm{W}x\ \&\ \mathrm{D}x)]$$

This situation can occur in two different ways, as shown in Fig. 2-5.

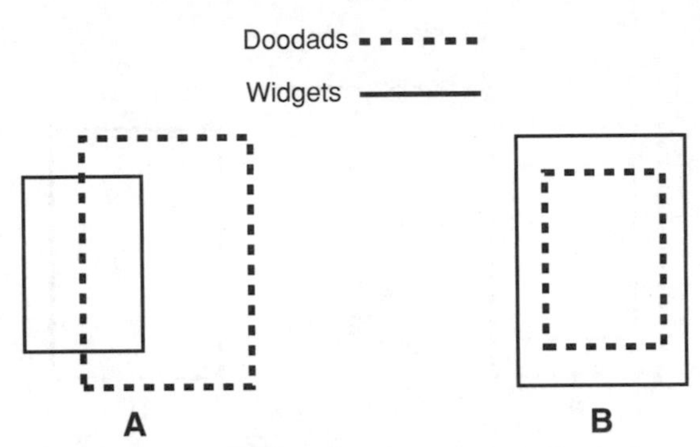

Fig. 2-5. Some (but not all) widgets are doodads. At A, the sets overlap but don't coincide. At B, the set of doodads is contained within the set of widgets.

NO WIDGETS ARE DOODADS

Suppose we want to illustrate this sentence: "No widgets are doodads." This can be more technically translated as, "It is not true that there exists an *x* such that *x* is a widget and *x* is a doodad." Symbolically, we write:

$$\neg[(\exists x)\ \mathrm{W}x\ \&\ \mathrm{D}x]$$

This situation can also be stated as "There exists no *x* such that *x* is a widget and *x* is a doodad." We symbolize this as above, but without the square brackets so the negation symbol applies directly to the quantifier:

$$\neg(\exists x)\ (\mathrm{W}x\ \&\ \mathrm{D}x)$$

This is illustrated by the Venn diagram of Fig. 2-6. The set of widgets and the set of doodads are *disjoint sets*. That means they have no elements (*x*'s, denoting widgets or doodads) in common. Another way of stating this is:

$$(\forall x)\ \neg(\mathrm{W}x\ \&\ \mathrm{D}x)$$

NOT ALL WIDGETS ARE DOODADS

Let's look at one more example. Suppose we want to illustrate this statement: "Not all widgets are doodads." This can be changed to the more rigorous form, "It is not true that for every *x*, if *x* is a widget, then *x* is a doodad." Symbolically, we get the following formula:

$$\neg[(\forall x)\ (\mathrm{W}x \Rightarrow \mathrm{D}x)]$$

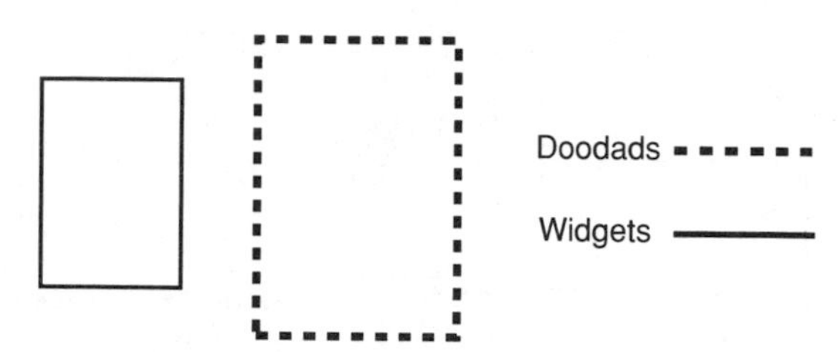

Fig. 2-6. No widgets are doodads. The sets are disjoint; they have no elements in common.

We can also say, "Not for every x, is it true that if x is a widget, then x is a doodad." This is written just the same as above, but without the square brackets. In this rendition, the negation sign applies directly to the quantifier:

$$\neg(\forall x)\ (\mathrm{W}x \Rightarrow \mathrm{D}x)$$

This is illustrated by means of the Venn diagrams in Fig. 2-7. Any imaginable scenario is possible, except those in which the set of widgets is a subset of, or is exactly the same as, the set of doodads.

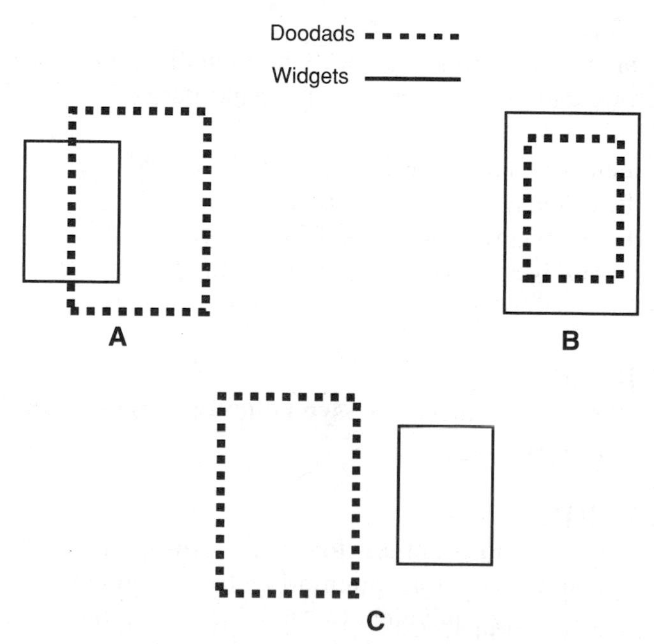

Fig. 2-7. Not all widgets are doodads. At A, the sets overlap but don't coincide. At B, the set of doodads is contained within the set of widgets. At C, the sets are disjoint.

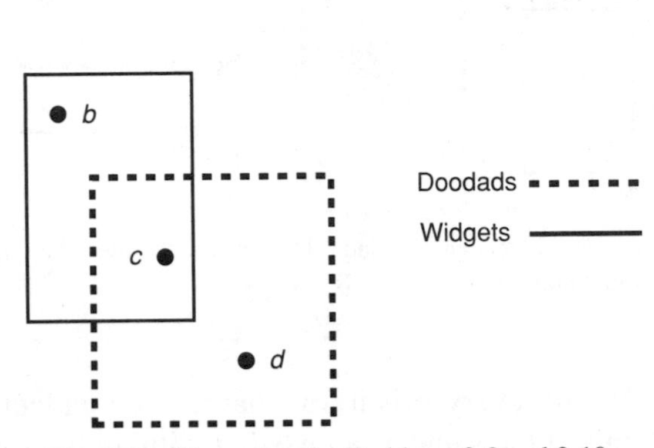

Fig. 2-8. Illustration for Problems 2-9 and 2-10.

PROBLEM 2-9

Examine Fig. 2-8. The set of widgets is shown as a solid rectangle, and the set of doodads is shown as a dashed rectangle. Four points are shown, representing constants *a*, *b*, *c*, and *d*. Write predicate sentences (in words) for all four of these constants.

SOLUTION 2-9

Here are the sentences that apply individually to each of the constants, which we call item *a*, item *b*, item *c*, and item *d*:

- Item *a* is not a widget and is not a doodad.
- Item *b* is a widget but not a doodad.
- Item *c* is a widget and is also a doodad.
- Item *d* is not a widget, but it is a doodad.

PROBLEM 2-10

Write the above sentences in symbolic form. Be sure they conform to the rules for wffs.

SOLUTION 2-10

Some people find it helpful to write word-based sentences in rigorous logical form before attempting to symbolize them. The word "but" is logically equivalent to "and." Here are the sentences in "wff-ready" form:

- It is not true that item *a* is a widget, and it is not true that item *a* is a doodad.
- It is true that item *b* is a widget, and it is not true that item *b* is a doodad.
- It is true that item *c* is a widget, and it is true that item *c* is a doodad.
- It is not true that item *d* is a widget, and it is true that item *d* is a doodad.

Based on these sentences, translation into wffs is straightforward. For any non-specified item *x*, let W*x* mean "*x* is a widget," and let D*x* mean "*x* is a doodad." The following four wffs are valid, based on Fig. 2-8:

$$\neg \mathrm{W}a \ \& \ \neg \mathrm{D}a$$
$$\mathrm{W}b \ \& \ \neg \mathrm{D}b$$
$$\mathrm{W}c \ \& \ \mathrm{D}c$$
$$\neg \mathrm{W}d \ \& \ \mathrm{D}d$$

Quiz

This is an "open book" quiz. You may refer to the text in this chapter. A good score is eight correct. Answers are in the back of the book.

1. Suppose we are given a sentence in symbolic form: $(\exists x)$ Px. The part of this sentence after the quantifier
   (a) is an SV sentence.
   (b) is an SVO sentence.
   (c) is an SLVC sentence.
   (d) might be an SV, SVO, or SLVC sentence; we don't know unless we are told what P stands for.

2. If Q is a sentence in propositional logic, then
   (a) Q is a wff.
   (b) Q is not a wff.
   (c) Q contains an existential quantifier.
   (d) Q contains a universal quantifier.

3. Suppose F, G, and H are complicated sentences, but all three are wffs. Which of the following is not a wff?
   (a) $\neg(F \,\&\, G) \Rightarrow H$
   (b) $\neg F \,\&\, \neg G \,\&\, \neg H$
   (c) $\neg F \neg\&G \neg\&H$
   (d) $F \vee G \vee \neg H$

4. Which of the following is an example of an SLVC sentence?
   (a) I know.
   (b) Jim runs to the Post Office.
   (c) We are prisoners.
   (d) Jane drives a truck.

5. Which of the following is an example of an SV sentence?
   (a) I know.
   (b) Jim runs to the Post Office.
   (c) We are prisoners.
   (d) Jane drives a truck.

6. Let the predicate symbol W stand for "is a widget," and let the predicate symbol D stand for "is a doodad." Imagine that there are lots of widgets and lots of doodads lying around. Let z be a variable. Suppose we know the following statement is true:

   $$(\forall z)\ (Wz \Rightarrow Dz)$$

   Based on this fact, of which of the following statements can we be certain?
   (a) $(\forall z)\ (Dz \Rightarrow Wz)$
   (b) $\neg(\forall z)\ (Dz \Rightarrow Wz)$
   (c) $(\exists z)\ (Wz \,\&\, Dz)$
   (d) All of the above

7. Consider the scenario of Question 6. Which of the Venn diagrams in Fig. 2-9 can apply to this situation?
   (a) A
   (b) B
   (c) C
   (d) None of the diagrams (A), (B), or (C) can apply.

8. In an SVO sentence, the subject is always
   (a) a noun.
   (b) a verb.

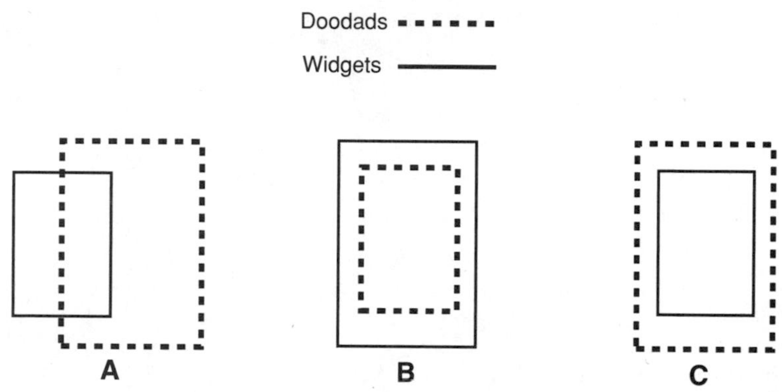

Fig. 2-9. Illustration for Quiz Question 7.

(c) an adjective.
(d) a wff.

9. A wff cannot contain
   (a) both negation and disjunction.
   (b) both negation and conjunction.
   (c) both constants and variables.
   (d) a variable all by itself, and nothing else.

10. Consider the statement "I created a TIFF image." The structure of this sentence can best be described as
    (a) SVO.
    (b) SV.
    (c) an existential quantifier.
    (d) a universal quantifier.

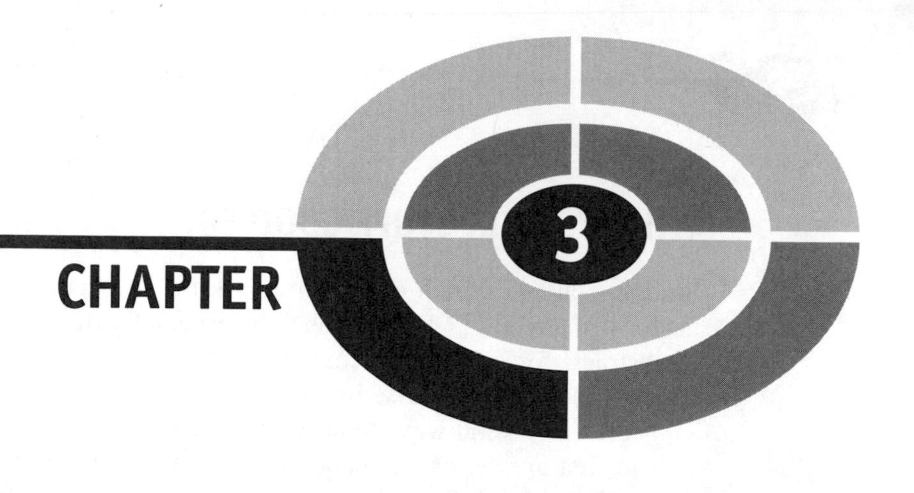

Formalities and Techniques

In this chapter, we'll look at the elements that make up a *mathematical theory*, also called a *mathematical system*. Then we'll examine the most common techniques used to prove propositions in a mathematical theory.

Seeds of a Theory

A mathematical theory is built up from certain initial assumptions, using definitions along with the rules of logic. The rules are used in an attempt to prove (or disprove) various statements known as *propositions*, which, once they have been proved, become *theorems*. These theorems can be used to prove (or disprove) other propositions.

WHAT IS THIS STUFF GOOD FOR?

Students taking a math course often ask the teacher or professor, "What's this useful for, anyhow?" This seems to be especially true in classes where the math is abstract: for example, a course in set theory, a course in number theory, a course in logic, or a course in how to prove theorems.

There are several ways to answer this question.

The first answer can be called the old-fashioned-country-schoolmaster reaction: "Never mind. Be quiet and study." That won't satisfy students today.

The second answer goes like this: "Don't worry about it. Someday you'll need this knowledge to solve a problem you couldn't solve otherwise. I can't tell you today what that problem will be. But you'll find out eventually, and then you will thank me." Maybe.

The third response is a series of counter-questions: "What do you mean by 'useful'? Making a faster airplane? Devising a better system to forecast the weather? Finding a cure for cancer? Getting rich? Having intellectual fun? Does everything have to be 'useful'? Can't knowledge and learning exist for their own sake?"

Pure mathematicians may have the best response to the question, "What is math useful for?" I recommend that you read *A Mathematician's Apology* by G. H. Hardy (Cambridge University Press, Cambridge, England, 1992). If your local library doesn't have it, you can get it through an online bookseller, such as *barnesandnoble.com* or *amazon.com*.

For now, let's not worry about practical matters. Let's dive into the business of math proofs, and forget all about the utilitarian world.

MATHEMATICAL RIGOR

In mathematics, the term *rigor* refers to the fact that a theory is built up methodically, based on logic. Nothing, except the few initial assumptions needed to get it going, is taken on faith. The term *rigor* in this context means "perfection, soundness, consistency, and elegance." A pure mathematician will tell you that a rigorous mathematical theory is very much alive—not in spite of the rigor, but because of it.

In a rigorous mathematical theory, every statement must be proved. If a statement is not proved, it cannot be taken as true. If a statement is proved false, then its negation is true.

A *contradiction* in a mathematical theory occurs when a statement P is proved, and the statement ¬P is also proved. When mathematicians develop theories, they hope contradictions do not arise. If a contradiction is found, it is the logical equivalent of what happens to a building when a demolition bomb is set off inside. Once a theory has thus been shown to be *logically unsound* or *logically inconsistent*, you can't trust a single theorem that was derived in it. Some of the provable theorems are true and others are not, but there is no way to know which theorems are good and which ones are bad. The logical "navigation compass" has been lost, and you are adrift in a sea of uncertainty.

DEFINITIONS

In order to know what we're talking about, we must define the basic terminology we intend to use when building a mathematical system. If you think a mathematical theory as a building, then definitions are like the stones or blocks of the foundation. Without a solid foundation, a building cannot be expected to withstand the tests to come. Without adequate definitions, a theory will not survive the scrutiny of thesis examiners, the challenge of applications to the real world, and all the other assaults it is bound to face.

With sound and sufficient definitions, flawless logic, and some good luck, a theory will evolve into something interesting and elegant, and might even make a major contribution to knowledge.

A definition often takes the form of an SLVC sentence. Sometimes it will contain an "if-and-only-if" statement. Here are some examples of rigorous definitions. Note the formal wording that resembles the language used in legal documents:

- A *set* is a collection or group of things called *elements*. A set is denoted by listing its elements in any order, and enclosing the list in curly brackets.
- The *empty set*, also called the *null set*, is the set containing no elements. It is symbolized { } or ∅. A set is empty if and only if it contains no elements.
- Let Q be a point in a flat, two-dimensional (2D) plane X. Let C be the set of all points in X that are at a distance r from Q, where r is not equal to 0. Then C comprises the *circle of radius r in X, centered at point Q*.
- Let Q be a point in a three-dimensional (3D) space Z. Let S be the set of all points in Z that are at a distance r from Q, where r is not equal to 0. Then S comprises the *sphere of radius r in Z, centered at point Q*.

Drawings can help portray some definitions, but in order to be truly rigorous, a definition must be precise and unambiguous, even without any illustrations. Fig. 3-1A shows an example of a circle in a flat plane as defined above. Fig. 3-1B shows an example of a sphere as defined above.

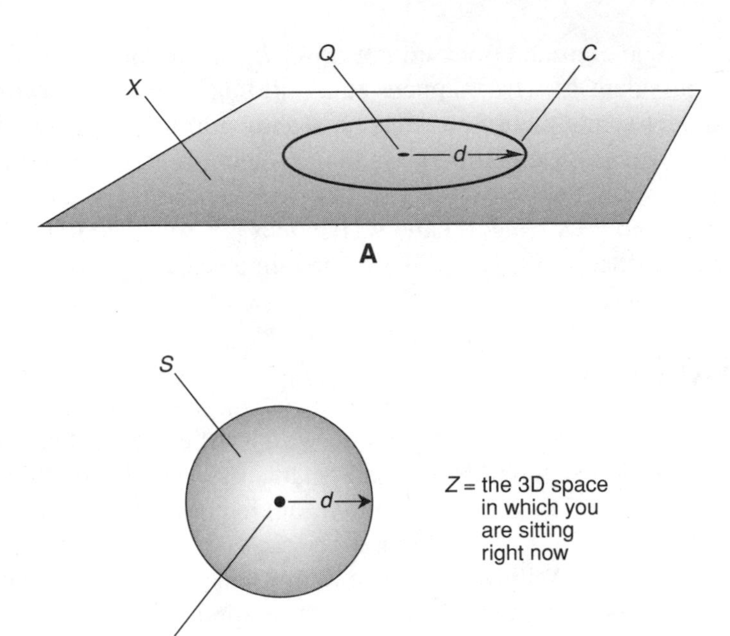

Fig. 3-1. At A, a circle in a plane. At B, a sphere in space.

A special sort of term, worthy of mention because it is just as important as standard defined terms, is the so-called *undefined term* or *elementary term*. Most mathematical theories have some of these. In geometry, the terms "point," "line," and "plane" are often considered elementary. They can be described so people have a good idea of what is meant by them, but they are difficult to rigorously define. In set theory, the notion of a "collection" or "group" is not rigorously defined. We can use synonyms like "bunch" or "few," "several" or "a lot," but the substitution of words alone does not make a definition.

AXIOMS

Along with definitions, a theory needs certain basic truths or assumptions with which to get started. We can't derive any true statements without at least a couple of things we accept on faith. In a mathematical theory, these assumptions are called *axioms* or *postulates*.

If definitions can be compared to the stones or blocks in the foundation of a building, then the axioms, taken all together, are analogous to the whole foundation. They are absolutely true by decree. Their truth holds, no matter what. When deciding on axioms to get a theory started, it's important that there be as few of them as possible, while still allowing for a theory that produces some meaningful results. If there aren't enough axioms, a substantial theory can't be built. If there are too many, the danger is high that a logical contradiction can be derived from them. Axioms usually (but not always) seem true to reasonable people.

Once we have plenty of defined terms and a few good axioms, logical rules can be applied, generating a larger group of truths. If the set of axioms is *logically consistent*, then a contradiction will never be encountered in the theory. When we decide on the axioms and write the definitions for a mathematical theory, we do not know for sure that a contradiction will never be found. But we minimize the risk by keeping the number of axioms to the minimum necessary so that the theory makes sense and produces enough theorems to be interesting.

Some classical examples of axioms follow. You might recognize these as the postulates set forth by the famous old-world mathematician, Euclid, when he developed his theory of geometry. Euclid's original wording has been changed slightly, in order to make the passages sound more contemporary. Let's not get sidetracked and rigorously define every term used here; you probably have a good idea already what they mean. (If not, you can consult a formal geometry textbook or even an online dictionary such as *www.dictionary.com*.)

- Any two points P and Q can be connected by a straight line segment (Fig. 3-2A).
- Any straight line segment can be extended indefinitely and continuously to form a straight line (Fig. 3-2B).
- Given any point P, a circle can be defined that has that point as its center and that has a specific radius r (Fig. 3-2C).
- All right angles are identical (Fig. 3-2D).
- Suppose two lines L and M lie in the same plane and both lines are crossed by the same straight line N. Suppose the measures of the adjacent interior angles x and y sum up to less than 180°. Then lines L and M intersect on the same side of line N as angles x and y are defined (Fig. 3-2E).

These are known as *Euclid's postulates*. As with definitions, a good axiom should not require an illustration to be clear and unambiguous. But there is no harm in making sketches if you find it helpful in the understanding of a principle, especially in geometry.

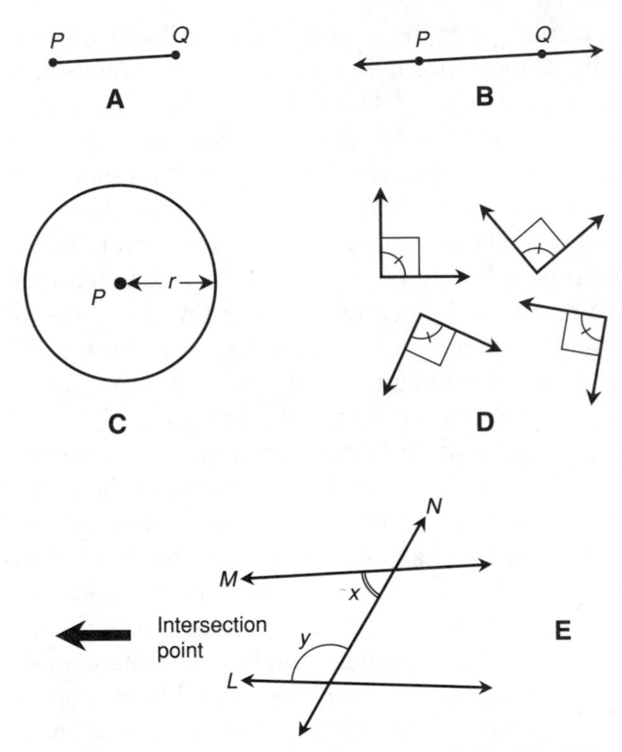

Fig. 3-2. Examples showing the concepts behind Euclid's original five axioms in classical geometry.

A DENIAL

The last axiom stated previously is known as *Euclid's Fifth Postulate*. It is logically equivalent to the following statement that has become known as the *Parallel Postulate*:

- Let L be a straight line, and let P be some point not on L. Then there exists one and only one straight line M, in the plane defined by line L and point P, that passes through point P and that is parallel to line L.

This axiom, and in particular its truth or untruth, has received enormous attention. If the Parallel Postulate is denied, the resulting system of geometry still works. It is consistent anyway! Here is its denial:

- Let L be a straight line, and let P be some point not on L. Then it is not true that there is one and only one straight line M, in the plane defined by line L and point P, that passes through point P and that is parallel to line L.

This means that in some "geometries" there is more than one line M, and in some "geometries" there are none. This is an example of a counter-intuitive axiom. But think about what happens when you try to draw parallel lines on the surface of the earth! They will always meet somewhere because the earth's surface is generally spherical. Lines of longitude are parallel at the equator, but they meet at the poles.

The people who first denied this postulate did so because they wondered what sort of mathematical theory would arise, and it is reasonable to suppose that they were prepared for the possibility that a contradiction would result. But it didn't, and the theories of *non-Euclidean geometry* were born.

PROPOSITIONS

Once a good set of definitions has been written up, and a set of axioms has been developed, we're ready to start building a mathematical theory.

Do you want to devise a new type of number system? A new way to think of geometry? Do you have an idea you'd like to pursue, such as the notion of numbers that can have more than one value? The process of theory-building involves taking the definitions and axioms and putting them together according to the rules of logic. Statements we intend to prove are called *propositions* until their truth has been firmly established.

Theorems

Once a proposition has been proved within the framework of a mathematical system, that proposition becomes a *theorem*. There are many well-known theorems in mathematics. If you've taken any geometry courses, you have learned a lot of theorems. If you're a mathematician, you can never get enough theorems, as long as no two of them contradict within the framework of a single mathematical system. (However, it is all right if a theorem that is true in one system is false in a different system.)

A CLASSICAL THEOREM

Here is an example of a theorem most people have heard or read by the time they graduate from high school:

- Let *A*, *B*, and *C* be three distinct points on a flat plane *X*. Consider the triangle formed by these points, symbolized ΔABC. Suppose that the sides of ΔABC are all straight, and they all lie in the flat plane determined by points *A*, *B*, and *C*. Let *a*, *b*, and *c* be the lengths of the sides of ΔABC opposite the points *A*, *B*, and *C*, respectively. Let the angle between the sides whose lengths are *a* and *b*, symbolized $\angle ACB$, be a right angle. Then the lengths of the sides of this triangle are related to each other according to the following equation:

$$a^2 + b^2 = c^2$$

Fig. 3-3 is a drawing that shows a right triangle with the points and sides labeled. The theorem has become known as the *Theorem of Pythagoras*, named after an old-world mathematician who, according to some stories, first proved it in the fifth century BC. It's also called the *Pythagorean Theorem*. (There is evidence, however, that the principle was known long before the time of Pythagoras.)

The Pythagorean Theorem is not an axiom. It was derived from the fundamental axioms and definitions of plane geometry, using the rules of logic. It is,

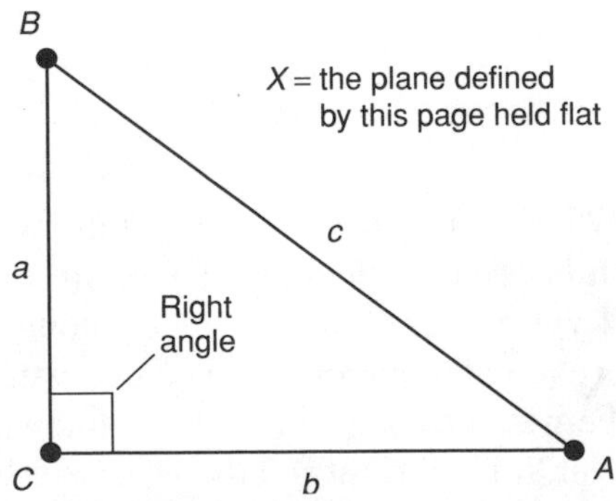

Fig. 3-3. The Pythagorean Theorem involves the relationship among the lengths of the sides of a right triangle. In this case, $a^2 + b^2 = c^2$.

in particular, dependent on the Parallel Postulate. The Pythagorean Theorem holds true only on flat surfaces. The equation does not work on spheres, such as the surface of the earth considered on a large scale. Nor does it work on conical surfaces, funnel-shaped surfaces, or saddle-shaped surfaces, or when any side of the triangle is not straight.

LEMMAS

Sometimes the proof of a theorem is simple and straightforward. In other cases the proof is complicated and lengthy. When the proof of a theorem involves many steps, it can help if we prove some preliminary theorems first, using them as stepping stones on the way to the final proof of the intended theorem. A minor theorem of this sort is called a *lemma*.

A lemma does not have to be of enduring interest, except for the fact that it plays a role in the construction of a major proof. But technically, a lemma is a theorem. When a lemma is proved, it can be saved for possible reuse in proving theorems to come. Once in a while a lemma will turn out, years or decades after it was first generated, to be more important than anybody thought!

COROLLARIES

Sometimes, when a theorem is proved, a few short steps can produce one or more other theorems. Such a secondary theorem is called a *corollary*.

Consider the Pythagorean Theorem equation. It holds true on a flat plane, but not on a curved surface. We might come up with a corollary to the Pythagorean Theorem that goes something like this:

- Let A, B, and C be three distinct points on a surface S. Consider the triangle formed by these points, symbolized ΔABC. Suppose that the sides of ΔABC all lie entirely on the surface S. Let a, b, and c be the lengths of the sides of ΔABC opposite the points A, B, and C, respectively. Let the angle between the sides whose lengths are a and b, symbolized $\angle ACB$, be a right angle. Suppose that the Pythagorean relationship, $a^2 + b^2 = c^2$, does not hold true. Then S is not a flat plane.

This can be proved from the Pythagorean Theorem using the law of implication reversal from propositional logic. In case you don't remember it, the law of implication reversal goes like this:

- Let X and Y be logical variables. Then the following logical statement is valid:

$$(X \Rightarrow Y) \Leftrightarrow (\neg Y \Rightarrow \neg X)$$

If we let X stand for "*S* is a flat surface" and Y stand for "The Pythagorean Theorem equation works on that surface," and if we let all other factors remain unchanged, the corollary can be proved by a single application of the law of implication reversal. The statement, "If a surface is flat, then the Pythagorean Theorem equation works on that surface" is logically equivalent to "If the Pythagorean Theorem equation doesn't work on a surface, then that surface isn't flat."

Figs. 3-4 and 3-5 show examples of right triangles on surfaces that are not flat. In Fig. 3-4, the surface is defined as *positively curved*; it bends in the same sense no matter what the orientation. In Fig. 3-5, the surface is defined as *negatively curved*; it bends in one sense for some orientations and in the opposite sense for other orientations. In neither case does the Pythagorean Theorem equation, in its traditional form, work for expressing the relationship among the lengths of the sides.

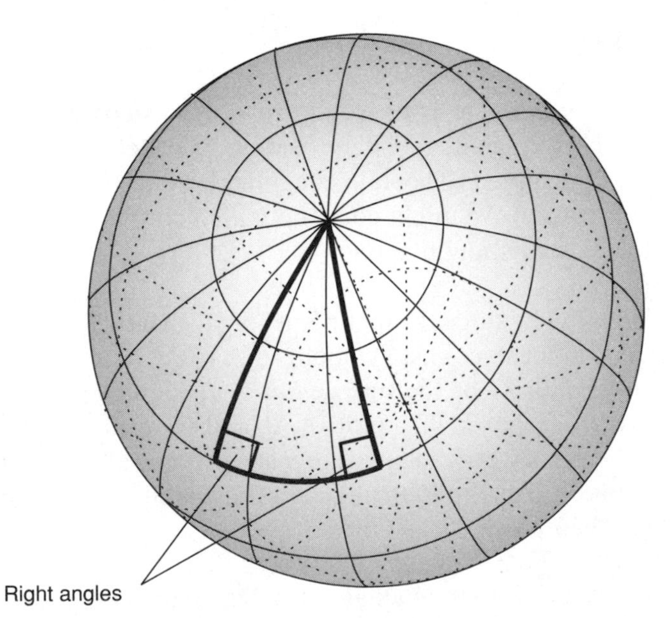

Fig. 3-4. A right triangle on a positively curved surface.

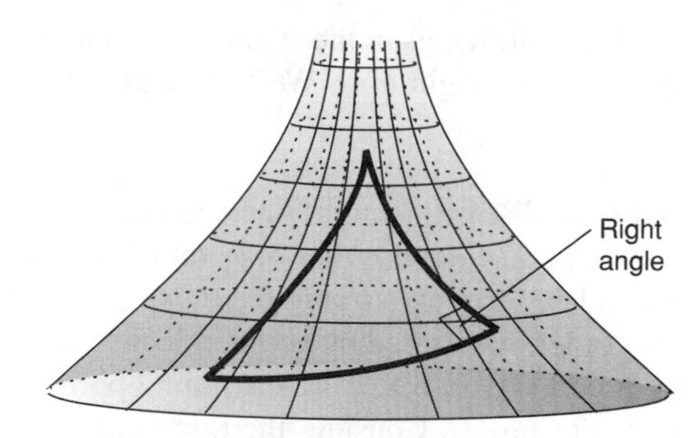

Fig. 3-5. A right triangle on a negatively curved surface.

PROBLEM 3-1

Define the meaning of the term *interior angle*, as it pertains to triangles.

SOLUTION 3-1

Let S, T, and U be specific, straight line segments. Suppose that S, T, and U are joined at their end points P, Q, and R. Consider the three angles $\angle PQR$, $\angle QRP$, and $\angle RPQ$. These three angles constitute the three *interior angles* of the triangle ΔPQR.

PROBLEM 3-2

Define the meaning of the term *measure of an angle in degrees*. Also provide symbology.

SOLUTION 3-2

Let S and T be specific, straight line segments. Let the end points of S be point Q and point R, and let the end points of T be point Q and point P, so line segments S and T intersect at their common end point Q. The *measure of angle* $\angle PQR$ *in degrees*, symbolized $m^\circ \angle PQR$, is the fraction of a circle that $\angle PQR$ subtends, multiplied by 360.

PROBLEM 3-3

A familiar theorem in geometry states that the interior angles of a triangle always add up to 180°, provided the triangle and its sides are entirely contained on a flat surface. Express this theorem formally.

Don't mention any drawings in the statement of the theorem. You don't have to supply a proof right now. We'll do that in Chapter 5.

SOLUTION 3-3

Let *S*, *T*, and *U* be distinct, straight line segments, all of which lie on a common flat plane *X*. Suppose that *S*, *T*, and *U* are joined at their end points *P*, *Q*, and *R*, forming a triangle ΔPQR. Suppose *S*, *T*, and *U* are named in that order by proceeding counterclockwise around ΔPQR. Let *P* be the point opposite side *S*, *Q* be the point opposite side *T*, and *R* be the point opposite side *U*. Consider the three interior angles of this triangle: $\angle PQR$, $\angle QRP$, and $\angle RPQ$. Then the following equation always holds true:

$$\text{m}^\circ\angle PQR + \text{m}^\circ\angle QRP + \text{m}^\circ\angle RPQ = 180^\circ$$

This isn't quite the way Euclid would have stated it, but it can suffice for us today.

PROBLEM 3-4

State a corollary to the preceding theorem that can be derived using the law of implication reversal. Again, don't mention any drawings in the statement of the corollary. We introduce a new definition here, as follows:

- On a surface *X*, a *geodesic* between two distinct points *A* and *B* is the line segment or curve *C* with end points *A* and *B*, such that *C* lies entirely on *X*, and such that *C* is shorter than any other line segment or curve on *X* whose end points are *A* and *B*.

SOLUTION 3-4

Let *S*, *T*, and *U* be distinct geodesics, all of which lie on a common surface *X*. Suppose that *S*, *T*, and *U* are joined at their end points *P*, *Q*, and *R*, forming a triangle ΔPQR. Suppose *S*, *T*, and *U* are named in that order by proceeding counterclockwise around ΔPQR. Let *P* be the point opposite side *S*, *Q* be the point opposite side *T*, and *R* be the point opposite side *U*. Consider the three interior angles of this triangle: $\angle PQR$, $\angle QRP$, and $\angle RPQ$. Suppose either of the following is true:

$$\text{m}^\circ\angle PQR + \text{m}^\circ\angle QRP + \text{m}^\circ\angle RPQ > 180^\circ$$

or

$$\text{m}^\circ\angle PQR + \text{m}^\circ\angle QRP + \text{m}^\circ\angle RPQ < 180^\circ$$

Then the surface *X*, on which ΔPQR lies, is not flat.

A Theory Grows

Once a few theorems have been proved, more propositions become tempting. They seem to suggest themselves, "pleading for proof." It is up to the mathematician to carry out the logical steps with the rigor and formality necessary to be sure the proof of a proposition is valid—if a proof can be found.

PROOFS AND TRUTH

Sometimes a proposition seems intuitively true, but a proof cannot be found. That can mean either of two things. Either the proposition really is not true, no matter how much it might seem so, or else the proposition is true but a proof is hard or impossible to find.

A mathematical theory is usually a *first-order logical system*. There are some truths in this kind of logical structure that cannot be proved. This strange fact was itself proved in 1930 by a German mathematician named Kurt Gödel. It practically caused a revolution in mathematics when it was published.

IT'S ART!

A good mathematical theory grows into a fascinating—one might even say artistic—structure of theorems and corollaries (Fig. 3-6). As long as a contradiction is not found, the theory may continue to grow for a long time, perhaps indefinitely. As the pure mathematician works on a new theory, he or she might have an application in mind. But a fascination with the subject, a creative urge, and pride in one's work are enough to keep the pure mathematician working.

According to G. H. Hardy, mathematical truths exist independent of human thought. All the facts are there. They would be there even if human beings ceased to exist, or even, for that matter, if there had never been any life on earth. It is up to us, should we be sufficiently curious, creative, and motivated, to seek them out.

THE FIRST GREAT FORMAL THEORY

Thousands of years ago, the Greek mathematician Euclid developed and published what some historians consider the first true axiomatic mathematical theory.

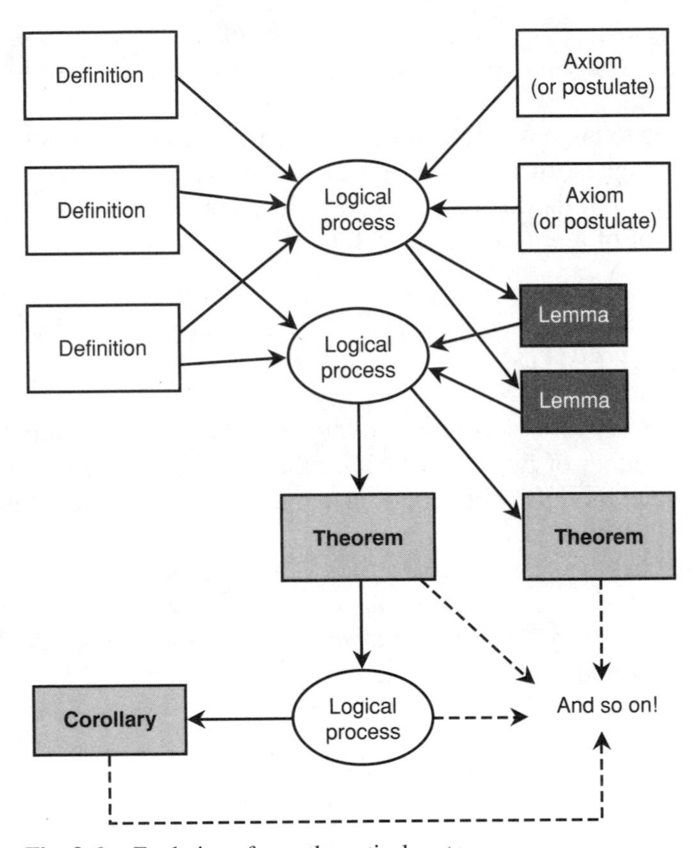

Fig. 3-6. Evolution of a mathematical system.

He proved that there exists an infinite number of *prime numbers*. The prime numbers are natural numbers (also called counting numbers) greater than 1, and cleanly divisible only by themselves and 1. But most people remember Euclid for his contributions to geometry. In particular, he practically invented plane geometry, which describes the relationships among points, lines, and figures on flat, two-dimensional surfaces.

Euclid began with several formal definitions and a set of postulates, including those stated earlier in this chapter. From these, he built up an elegant and far-reaching theory that we now know as *Euclidean Geometry*. A variant of it is still used today by surveyors and architects, and is still taught in schools all over the world.

For a good presentation of Euclid's original theory of geometry, you can get a copy of *The Elements* (Green Lion Press, Santa Fe, NM, 2002) from an online bookseller, such as *barnesandnoble.com* or *amazon.com*. Look for ISBN 1888009195. You might also find it at your local book shop or at a public or school library.

Techniques for Proving Things

There are a few powerful tactics that mathematicians use to prove theorems. All of the processes are based on the rules of logic.

DEDUCTIVE REASONING

Deductive reasoning, also called *deduction*, is the most straightforward means of proving theorems. The expression "deduction" should not be confused with the process of elimination that is sometimes used to argue in favor of something by discounting all the alternatives. In mathematics, deduction requires the use of logical rules in discrete steps, such as the law of implication reversal, the law of double negation, the distributive laws, and DeMorgan's laws, along with a few rules for predicate logic.

Consider the following, stated as a proposition:

- Everyone who lives in Wyoming likes beef. Joe lives in Wyoming. Therefore, Joe likes beef.

We can symbolize this as follows:

$$\{[\forall x\ (\mathrm{W}x \Rightarrow \mathrm{B}x)]\ \&\ \mathrm{W}j\} \Rightarrow \mathrm{B}j$$

In this wff, the predicate W means "lives in Wyoming." The predicate B means "likes beef." The symbol x is a variable (that is, a nonspecific individual) in the set of human beings. The symbol j is a constant—in this case, a certain person named Joe. The upside-down A is the universal quantifier, read "For every" or "For all."

Here is another popular way to write down this argument in symbolic form, emphasizing the step-by-step nature of it:

$\forall x\ (Wx \Rightarrow Bx)$
Wj
$\therefore$
Bj

In this set of statements, progressive facts are listed one below the other. The three-dot symbol means "therefore."

There's yet another way this formula can be written:

$\forall x\ (Wx \Rightarrow Bx)$
Wj

Bj

In this layout, the three dots are replaced by a horizontal bar, so the argument looks a little like a sum in arithmetic.

This is a good example of deductive reasoning. It follows from the fact that if something is true for a variable in a given set, then it is true for any constant (that is, any individual) in that set. This fact is a known theorem of predicate logic, and has been proved within the framework of that discipline.

LEAP BEFORE YOU THINK FURTHER!

Let's take for granted everything that has ever been proved in propositional and predicate logic, and consider all these theorems as legitimate tools in our arsenal for proving propositions in all other disciplines. That way, we don't have to go through the proofs of all the known theorems in all of propositional and predicate logic, a process that would itself require its own book. One of the main reasons logic was developed was to make it possible to prove theorems in other mathematical fields.

A roster of useful theorems from propositional logic was presented in Chapter 1. Proofs of these theorems are fairly easy because the "brute force" tactic of truth tables can be used. For predicate logic, the situation becomes more complicated.

WHAT'S THE UNIVERSE?

Before you try to prove propositions and thereby make them into theorems, it's a big help if you know what you are trying to prove things about. Are you trying to prove a proposition involving the set of human beings? Are you trying to prove things about numbers? Planets? Triangles? Spheres? What?

Sometimes the *universal set*, also called the *universe*, is the set of all objects to which a theorem is intended to apply. This should be evident from context. For example, consider this statement: "If x is a real number, then $x + 1$ is also a real number." The universe is the set of all real numbers. Or this: "The sum of the measures of the interior angles of any triangle on a flat surface is 180°." The universe is the set of triangles on flat surfaces.

If the universal set is not evident from context, then the universal set should be plainly stated. For example, we can preface a proposition with a sentence such as this: "Let x be a real number." Or this: "Let T be a triangle on a flat surface."

TO PROVE "THERE EXISTS..."

Suppose that you want to prove that a proposition holds true for some, but not necessarily all, of the objects within a universal set U. Imagine that you are confronted with this proposition:

- Let W be the set of all widgets. Let D be the set of all doodads. Then there exists some element w in set W, such that w is an element of set D.

In set theory, the phrase "is an element of" is symbolized $\in$. This looks a little like the Greek letter epsilon, or a mutated English uppercase letter E. Given this symbol, we can write the above theorem as follows, based on the knowledge of what the letters stand for:

$$(\exists w)\ [(w \in W)\ \&\ (w \in D)]$$

In order to prove a proposition of this form, we only need to provide one example for which the statement is true. Once we have shown that this type of proposition holds true for one object, we have shown that it holds true "for some" objects. That is, "there exists" an object for which the proposition is true.

Theorems of this sort can be called *weak theorems*. They don't always tell us much. But they can sometimes be significant.

TO PROVE "FOR ALL..."

In mathematics, it's common to come across a proposition that claims a fact for all objects within a universe U. These propositions, once proved, can be called *strong theorems*. We stated a couple of examples a while ago. Here they are again, written down in more formal style.

- Let $\Re$ represent the set of all real numbers. For all x, if x is an element of $\Re$, then $(x + 1)$ is an element of $\Re$.

Symbolically:

$$(\forall x) \{(x \in \Re) \Rightarrow [(x + 1) \in \Re]\}$$

- Let T represent the set of all triangles on flat surfaces. Let H represent the set of all triangles for which the sum of the measures of the interior angles is 180°. For all y, if y is an element of T, then y is an element of H.

Symbolically:

$$(\forall y) [(y \in T) \Rightarrow (y \in H)]$$

In order to prove a proposition of this form, we have to show that for any arbitrary (that is, nonspecific) object, the proposition holds true. It is not good enough to show that the proposition holds true in specific cases, even a large number of them. We have to find a way to demonstrate that the proposition is a fact for all of the objects in the universal set, without fail, even if the universal set has an infinite number of elements.

One of the most amazing things about mathematics is that it's possible to prove something for infinitely many things while executing only a finite number of steps. Sometimes this is easy, and sometimes it is difficult. Sometimes, insight and intuition come into play in ways more often associated with artists than technicians.

INSTANCES OF "FOR ALL..."

Consider the following statement: "All rational numbers are real numbers." Suppose we are assured that this statement is true. Now imagine that we are interested in the properties of the number −577/843. Is this a real number? If we are able to show that −577/843 is a rational number, then according to the statement given above, we can conclude that it is a real number. (If we are not able to show that −577/843 is a rational number, it doesn't prove that −577/843 is not a real number; it merely tells us nothing important.)

If you have taken any middle-school mathematics courses, you should know what a rational number is. Here is a formal definition:

- Let x be a number. Then x is a *rational number* if and only if x can be expressed in the form a/b, where all of the following are true: (1) a is an integer, and (2) b is a natural number, and (3) $b \neq 0$ (b is not equal to 0).

In case you've forgotten what natural numbers and integers are, here are formal definitions for them:

- Let b be a number. Then b is a *natural number* if and only if b is an element of the set $N = \{0, 1, 2, 3, 4,\ldots\}$.
- Let a be a number. Then a is an *integer* if and only if one of the following is true: (1) a is a natural number, or (2) $-a$ is a natural number.

Let's consider *number* an elementary term for the purposes of this discussion. The following can serve as an informal definition of a real number:

- Let x be a number. Then x is a *real number* if and only if one of the following is true: (1) $x = 0$, or (2) x can be used to express the distance in specific units between two geometric points, or (3) $-x$ can be used to express the distance in specific units between two geometric points.

It's clear that −577/843 is of the form a/b. All we have to do is let $a = -577$ and $b = 843$. It happens to be true that the negative of −577 is 577, so a is an integer. It also happens to be true that 843 is a natural number, because $843 \in \{0, 1, 2, 3, 4,\ldots\}$. It is obvious that 843 is not equal to 0. Therefore, according to the original proposition, −577/843 is a rational number. Because all rationals are reals, −577/843 is a real number.

Let the predicate Q stand for "is a rational number." Let the predicate R stand for "is a real number." Let x be a logical variable. Let k be the constant −577/843. Then we can write our single-instance proof like this:

$$\forall(x)\ \mathrm{Q}x \Rightarrow \mathrm{R}x$$
$$\mathrm{Q}k$$

$$\mathrm{R}k$$

INSTANCES OF "THERE EXISTS..."

Consider this statement: "Some natural numbers can be divided by 7, and the result is another natural number." Suppose we are interested in the number 765. This is a natural number. Can it be divided by 7 to get another natural number? There is only one way to find out: test it and see. A calculator can be used to do this. Divide 765 by 7, and see what you get! If your calculator agrees with mine, then you should get:

$$765/7 = 109.285714285714285714\ldots$$

That quotient is not a natural number.

Let's test another number, this time 322. If your calculator agrees with mine, then you should get:

$$322/7 = 46$$

That quotient is a natural number.

When you want to demonstrate that a certain constant satisfies an existential proposition (that is, one of the form "There exists..."), all you have to do is test it and hope that the test comes out positive. Of course, if the test fails, you have proved exactly the opposite: that the constant does not satisfy the proposition. But you have not disproved the entire proposition. In fact, disproving an existential proposition can be difficult. To do that, you have to prove that the proposition never holds true.

REDUCTIO AD ABSURDUM

One of the most elegant and interesting tactics in the mathematician's arsenal is known as *reductio ad absurdum*. This is Latin and translates roughly to "to reduce to absurdity." In order to use this technique, we start by assuming that what we want to prove is false! Then from this, we derive a contradiction. That proves that our assumption is false, so the original proposition must be true.

Some purists argue that *reductio ad absurdum* should be used only when all other attempts at proof have failed. But there are situations that seem to cry out for the use of this technique. In particular, statements of the form "There exist no..." are ideal candidates.

We've defined rational numbers, and we have a good idea of what a real number is. There are real numbers that are not rational numbers. They are defined this way:

- Let x be a number. Then x is an *irrational number* if and only if both of the following are true: (1) x is a real number, and (2) x is not a rational number.

Now consider this proposition:

- No irrational number can be expressed as an integer divided by a nonzero natural number.

In order to prove it, let's assume its opposite. Call this assumption A. We will try to derive a contradiction from A. This assumption A, which is the negation of the original proposition, is:

- A = There is at least one irrational number that can be expressed as an integer divided by a nonzero natural number.

Suppose that y is such a number. Then $y = a/b$, where a is an integer and b is a nonzero natural number. It follows that y is rational, because it meets the definition of a rational number. But we just got done specifying that y is irrational! We have a direct contradiction; we are saying that y is a rational number, and y is not a rational number. We must conclude that the assumption A is false, and therefore that ¬A is true. That proves the following:

- It is not true that there is at least one irrational number that can be expressed as an integer divided by a nonzero natural number.

This is logically identical to the original proposition, which we can state again to be sure we're not getting confused by all the quantifiers and negations:

- No irrational number can be expressed as an integer divided by a nonzero natural number.

PROBLEM 3-5

Use *reductio ad absurdum* to show that there exists no largest rational number.

SOLUTION 3-5

Suppose there is a largest rational number. Call it r. That is our assumption, which we will call A. Let's prove that A is false.

According to the definition of rational number, $r = a/b$, where a is some integer and b is some nonzero natural number. We can be certain that $r > 0$, because all the positive rational numbers are larger than any negative rational number or 0. This means that a is a positive integer, that is, $a > 0$. (If a were negative, then r would be negative, and if a were equal to 0, then r would be equal to 0, and we've ruled those possibilities out.)

Now consider a number s such that the following is true:

$$s = (a + 1)/b$$

We can be certain that $a + 1$ is an integer, because 1 plus any integer always equals another integer. This means that $(a + 1)/b$ is a rational number, and therefore that s is a rational number. We also know that $(a + 1) > a$. Because a is positive, $(a + 1)/b > a/b$. That means $s > a/b$, and therefore that $s > r$. Thus s is rational, and $s > r$. This contradicts our assumption A, that there is a largest rational number, so A must be false. That means ¬A is true: There exists no largest rational number.

MATHEMATICAL INDUCTION

There's one more technique that we'll look at here. It is known as *mathematical induction*. A less formal term might be *proof by mathematical chain reaction*. Using this scheme, it is possible to prove things about all the elements in an infinite set, using only a finite number of steps.

Imagine an infinite set S, consisting of elements called a_0, a_1, a_2, a_3, a_4, and so on, like this:

$$S = \{a_0, a_1, a_2, a_3, a_4, \ldots\}$$

Suppose that we want to prove that a proposition P is true about all the elements of S. We can't prove P for each element of S one by one, or even for large batches of elements, because the list goes on forever. But suppose we can prove that P is true for a_0, the first element in S. Also suppose we can prove that if P is true for some unspecified element a_n in set S (where n is a natural number), then P is true for the next element $a_{(n+1)}$ in set S. By doing these two things, we create a "chain reaction of truths." We know P is true for the first element, and this proves that P is true for the second; that in turn proves P for the third; and so it goes on forever, like an infinitely long line of dominoes knocking each other down.

PROBLEM 3-6

Show that for any two distinct rational numbers, there is a third rational number whose value lies between them. Don't use *reductio ad absurdum*, and don't try to use mathematical induction. It is all right, however, to take all the general rules of arithmetic (sums, products, differences, and quotients) for granted.

SOLUTION 3-6

Let the two rational numbers in question be called r and s. Suppose that the following are true:

$$r = a/b$$
$$s = c/d$$

where a and c are integers, and b and d are nonzero natural numbers. We know such numbers a, b, c, and d exist, because r and s are both rational, and the definition of rational number requires that there exist such numbers a, b, c, and d.

Now consider the mathematical average of the two numbers r and s. We know this number lies between r and s, because the average (or arithmetic mean) of any two numbers is always between them (midway between, in fact!). Call this number x. If we can prove that x is rational, then we have proved the theorem.

We know the following about x, from the formula for finding averages:

$$x = (r + s)/2$$

That means that this general equation holds true for x:

$$x = (a/b + c/d)/2$$

From arithmetic, remember the general formula for sums of quotients:

$$a/b + c/d = (ad + bc)/bd$$

Therefore, we know that the following is true:

$$(a/b + c/d)\ /\ 2 = (ad + bc)/(2bd)$$

and thus:

$$x = (ad + bc)/(2bd)$$

If we can show that the quantity $(ad + bc)$ is an integer and that the quantity $2bd$ is a nonzero natural number, then we have shown that x is rational. The product of any two integers is always an integer; therefore ad and bc are integers. The sum of any two integers is always an integer; therefore the quantity $(ad + bc)$ is an integer. The product of any two nonzero natural numbers is a nonzero natural number; therefore bd is a nonzero natural number. Twice any nonzero natural number is a nonzero natural number; therefore the quantity $2bd$ is a nonzero natural number. All of this demonstrates that x is equal to an integer divided by a nonzero natural number, and therefore that x is rational. As previously stated, x is the arithmetic mean of r and s, so x lies between r and s. Therefore, for any two distinct rational numbers, there is a third rational number whose value lies between them.

PROBLEM 3-7

Use mathematical induction to show that all natural-number multiples of 0.1 are rational numbers.

SOLUTION 3-7

Remember that the set N of natural numbers is:

$$N = \{0, 1, 2, 3, 4, \ldots\}$$

Therefore, the set M of natural-number multiples of 0.1 is:

$$\begin{aligned} M &= \{(0 \times 0.1), (1 \times 0.1), (2 \times 0.1), (3 \times 0.1), (4 \times 0.1), \ldots\} \\ &= \{0, 0.1, 0.2, 0.3, 0.4, \ldots\} \end{aligned}$$

The first element of this set, 0, is rational, because it can be expressed in the form a/b, where a is an integer and b is a nonzero natural number. Simply let $a = 0$ and $b = 1$. This is the easy part of the induction proof.

Now for the hard part. Suppose that $n \times 0.1$ (which can also be written as $0.1n$) is rational for some unspecified natural number n. Consider the next number in our set M of multiples, $(n + 1) \times 0.1$. This can be rearranged using the rules of arithmetic:

$$\begin{aligned}(n + 1) \times 0.1 &= (n \times 0.1) + (1 \times 0.1)\\ &= 0.1n + 0.1\end{aligned}$$

We know that there exists some integer a and some nonzero natural number b such that $0.1n = a/b$, because we are given that $0.1n$ is rational. Therefore, we can rewrite the above expression as:

$$\begin{aligned}0.1n + 0.1 &= a/b + 0.1\\ &= a/b + 1/10\end{aligned}$$

Using the arithmetic rule for the sum of two quotients, we can rearrange the above as follows:

$$a/b + 1/10 = (10a + b)/10b$$

Ten times any integer is another integer; this is a known rule of arithmetic. Therefore, $10a$ is an integer. The sum of any integer and a nonzero natural number is an integer; this is another rule of arithmetic. Therefore, $10a + b$ is an integer. Ten times any nonzero natural number is another nonzero natural number; this is yet another rule of arithmetic. Therefore, the quantity $(10a + b)/10b$ is equal to an integer divided by a nonzero natural number. This means, by definition, that $(10a + b)/10b$ is rational. It also happens to be the same quantity as $a/b + 1/10$, which in turn is equal to $0.1n + 0.1$, the element immediately after $0.1n$ in the set M.

We have just proved that if any unspecified element of M is rational, then the next element is rational as well. That, in addition to the proof that the first element in M is rational, is all we need to claim that every element in the set M is rational, based on the principle of mathematical induction.

Quiz

This is an "open book" quiz. You may refer to the text in this chapter. A good score is eight correct. Answers are in the back of the book.

1. Imagine that you want to create an entirely new mathematical theory. The number of postulates in your theory
   (a) can be unlimited, and the more the better.
   (b) should be large enough so that a contradiction will be easy to derive.
   (c) should be as small as possible, while still producing a meaningful theory.
   (d) should be zero. You should never assume anything without proof.

2. Consider the following series of statements:

   $(\forall x)(Kx \Rightarrow Rx)$
   Kg
   ―――――――――
   Rg

   This is a generic symbolization of a proof by means of
   (a) *reductio ad absurdum*.
   (b) mathematical induction.
   (c) elementary terminology.
   (d) straightforward logical deduction.

3. Which of the following symbols is used to denote the fact that an object is an element of a particular set?
   (a) $\cup$
   (b) $\cap$
   (c) $\in$
   (d) $\not\subset$

4. Something that can be described so we have a good idea of what it means, but that is not rigorously defined, is called
   (a) an elementary term.
   (b) an axiom.
   (c) a postulate.
   (d) a lemma.

5. An axiom or postulate is
   (a) a fact proved on the basis of other known facts.
   (b) something assumed to be true without proof.
   (c) a minor theorem that follows easily from the proof of a major theorem.
   (d) a major theorem that is used to prove a minor theorem.

6. Consider the following series of statements:

   $\neg D \Rightarrow (H \,\&\, \neg H)$

   D

   This is a generic symbolization of a proof by means of

   (a) *reductio ad absurdum.*
   (b) mathematical induction.
   (c) elementary terminology.
   (d) straightforward logical deduction.

7. A mathematical theory can be rendered completely invalid if it is possible to prove, based on its axioms and definitions,

   (a) an infinite number of propositions.
   (b) only propositions containing existential quantifiers.
   (c) only propositions containing universal quantifiers.
   (d) a proposition and also its negation.

8. The set of all objects to which a theorem is intended to apply is called

   (a) the propositional set.
   (b) the predicate set.
   (c) the empty set.
   (d) the universe.

9. Suppose you are told that a certain proposition P holds true for some, but not all, rational numbers. You want to prove that P is true for 589/777. The most straightforward, and probably the easiest, way to do this is to demonstrate that P holds true for

   (a) 589/777.
   (b) all the positive rational numbers.
   (c) all the rational numbers between 0 and 1.
   (d) all the real numbers.

10. Fill in the blank in the following sentence to make it true: "If a proposition P holds true in general for a variable x in a set S, then P is true for any ________ in the set S."

    (a) constant
    (b) axiom
    (c) definition
    (d) corollary

Vagaries of Logic

History is replete with good theorems whose original proofs contained errors, and bad theorems that collapsed when their proofs were shown to be invalid. In this chapter we'll examine some classical *fallacies*, which are violations or misapplications of the laws of reason. We'll also look at a few *paradoxes*, which are incredible results that arise from seemingly sound arguments.

Cause, Effect, and Implication

When two things are correlated, it's tempting to conclude that there is a cause-and-effect relationship involved. Examples of this sort of flawed thinking abound. Anyone who listens to the radio, reads newspapers, or watches television can't escape them. Sometimes, a dubious cause-effect relationship is not directly stated, but only implied. "Take this pill and you'll be happy all the time. Avoid these foods and you won't die of a heart attack."

Imagine what some of the advertisements, and the weird logic they contain, sound like to people who have grown up in cultures radically different from ours! "If you drink this fizzy liquid after eating abominable food, then all of your abdominal pains will go away." "If you eat this purple goo, then all the ugly hairs in your nose will fall out." The implications vary, but the general nature of the implications is always the same: "If you do something that causes me to make money, then you will become happy." How do we know if there really is a cause-effect relationship when we see two things happen at the same time, or one right after the other?

CORRELATION AND CAUSATION

Suppose two phenomena, called *X* and *Y*, vary in intensity with time. Fig. 4-1 shows a relative graph of the variations in both phenomena. The phenomena shown in this graph change in a manner that is positively correlated. When *X* increases, so does *Y*, in general. When *Y* decreases, so does *X*, in general. (Negative correlation can also exist, where an increase in one factor is attended by a decrease in the other.) The *independent variable* is some factor that does not depend on either *X* or *Y*. Time is a common example of an independent variable.

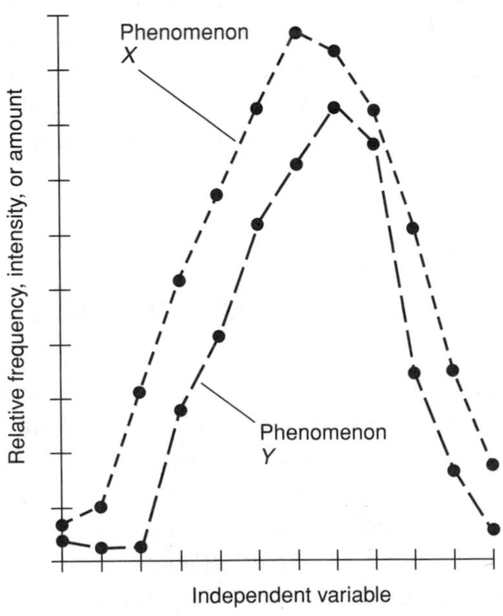

Fig. 4-1. The two phenomena shown here, *X* and *Y*, appear to be strongly correlated. Does correlation imply causation?

Is causation involved in the situation shown by Fig. 4-1? Maybe! There are four ways that causation can exist. But maybe there is no cause-and-effect relationship. If the scenario portrayed by Fig. 4-1 is that sort of case, then it shows nothing more than a *coincidence*: the phenomena *X* and *Y* closely coincide, but there is no particular reason why.

If there were 1,000 points on each plot, and they still followed each other the way they do in Fig. 4-1, there would be a better case for believing that causation is involved. As it is, there are only 12 points on each plot. Suppose these points represent a freak scenario? Or, suppose the 12 points in each plot of Fig. 4-1 have been selected by someone with a vested interest in the outcome of the analysis?

X CAUSES *Y*

Cause-and-effect relationships can be illustrated using arrows. Fig. 4-2A shows the situation where changes in phenomenon *X* directly cause changes in phenomenon *Y*. You can doubtless think of some scenarios. Here's a good real-life example.

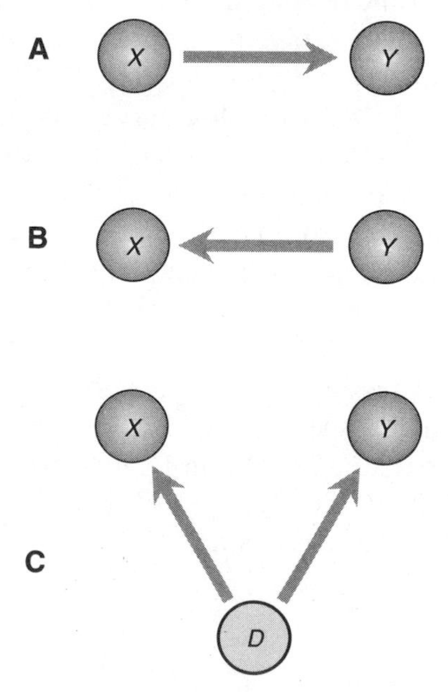

Fig. 4-2. At A, *X* causes *Y*. At B, *Y* causes *X*. At C, *D* causes both *X* and *Y*.

Suppose the independent variable, shown on the horizontal axis in Fig. 4-1, is the time of day between sunrise and sunset. Plot *X* shows the relative intensity of sunshine during this time period; plot *Y* shows the relative temperature over that same period of time. We can argue that the brilliance of the sunshine causes the changes in temperature. There is some time lag in the temperature function; this is to be expected. The hottest part of the day is usually a little later than the time when the sunshine is most direct.

It's harder to believe that there's a cause-and-effect relationship in the other direction. It is silly to suggest that temperature changes cause differences in the brilliance of the sunlight reaching the earth's surface. Right? Well, think some more! Suppose heating of the atmosphere causes clouds to dissipate, resulting in more sunlight reaching the surface (*Y* causes *X*)? Maybe we're looking at a situation where *Y* causes *X*.

Y CAUSES *X*

Imagine that the horizontal axis in Fig. 4-1 represents 12 different groups of people in a medical research survey. Each hash mark on the horizontal axis represents one group. Plot *X* is a point-to-point graph of the relative number of fatal strokes in a given year for the people in each of the 12 groups; plot *Y* is a point-to-point graph of the relative average blood pressure levels of the people in the 12 groups during the same year. (These are hypothetical graphs, not based on real historical experiments, but a real-life survey might come up with results something like this. Medical research has shown a correlation between blood pressure and the frequency of fatal strokes.)

Is there a cause-effect relationship between the value of *X* and the value of *Y* here? Most doctors would answer with a cautious "Yes." Variations in *Y* cause, or at least contribute to, observed variations in *X* (Fig. 4-2B). High blood pressure can be a cause of fatal strokes, in the sense that, if all other factors are equal, a person with high blood pressure is more likely to have a fatal stroke than a person identical in every other respect, but with normal blood pressure.

What about the reverse argument? Can fatal strokes cause high blood pressure (*X* causes *Y*)? No. That's clearly impossible. Once a person has had a fatal stroke, his or her blood pressure drops fast!

COMPLICATIONS

If you are a weather expert or a doctor and you are reading this, are you getting a little nervous? The above scenarios are oversimplified. The cause-and-effect relationships described here aren't "pure." In real life, events rarely occur with a single clear-cut cause and a single inevitable effect.

The brightness of sunshine is not, all by itself, the only cause of changes in the temperature during the course of a day. A nearby lake or ocean, the wind direction and speed, and the passage of a weather front can all have an effect on the temperature at any given location. We've all seen the weather clear and brighten, along with an abrupt drop in temperature, after a strong cold front passes by. The sun comes out, and it gets cooler. That defies the notion that bright sun causes things to heat up, even though the notion, in its "pure" form where all other factors are equal, is valid. The trouble is that other factors are not always equal!

In regards to the blood-pressure-versus-stroke relationship, there are numerous other factors involved, and scientists aren't sure they know them all. New discoveries are constantly being made in this field. Examples of other factors that are believed to play cause-effect roles in the occurrence of fatal strokes include nutrition, stress, cholesterol level, body fat index, presence or absence of diabetes, age, and heredity. A cause-effect relationship (Y causes X) exists between blood pressure and fatal strokes, but this relationship is not "pure."

D CAUSES BOTH *X* AND *Y*

Suppose that the horizontal axis in Fig. 4-1 represents 12 different groups of people in another medical research survey. Again, each hash mark on the horizontal axis represents one group. Plot X is a point-to-point graph of the relative number of heart attacks in a given year for the people in each of the 12 groups; plot Y is a point-to-point graph of the relative average blood cholesterol levels of the people in the 12 groups during the same year. As in the stroke scenario, these are hypothetical graphs. But they're plausible. Medicine has shown a correlation between blood cholesterol and the frequency of heart attacks.

Before I make enemies in the medical profession or the food industry, let me say that the purpose of this discussion is not to resolve the cholesterol-versus-heart-disease issue, but to illustrate complex cause-effect relationships. It's easier to understand a discussion about real-life factors than to leave things entirely generic. I do not have the answer to the cholesterol-versus-heart-disease riddle. If I did, I'd be writing a different book.

When scientists first began examine the hearts of people who died of heart attacks in the early and middle 1900s, they found "lumps" called *plaques* in the arteries. It was theorized that plaques cause the blood flow to slow down, contributing to clots that eventually cut off the blood to part of the heart, causing tissue death. The plaques were found to contain cholesterol. Evidently, cholesterol can accumulate inside the arteries. When they saw data showing a correlation between blood cholesterol levels and heart attacks, scientists got the idea that if the level of cholesterol in the blood could be reduced, the likelihood of the person having a heart attack later in life would go down. The theory was that fewer or smaller plaques would form, reducing the chances that clot formation could obstruct an artery.

One of the first thing doctors did after the correlation was found was to tell heart patients to reduce the amount of cholesterol-containing foods in their diet, hoping that this change in eating behavior would cause blood cholesterol levels to go down. In many cases, a low-cholesterol diet did bring down blood cholesterol levels. (Later, drugs were developed that had the same effect.) Studies continue along these lines to this very day, and will continue for years to come. It is becoming apparent that reducing the amount of cholesterol in the diet, mainly by substituting fruits, vegetables, and whole grains for cholesterol-rich foods, can reduce the levels of cholesterol in the blood. This type of dietary improvement can apparently also reduce the likelihood that a person will have a heart attack later in life. There's more than mere correlation going on here. There's causation, too. But how much causation is there? Between what variables, and in what directions, does it operate?

Let's call the amount of dietary cholesterol by the name factor D. According to current popular medical theory, there is a cause-and-effect relation between this factor and both X and Y. Some studies have indicated that, all other things being equal, people who eat lots of cholesterol-rich foods have more heart attacks than people whose diets are cholesterol-lean. The scenario is shown in Fig. 4-2C. There is a cause-and-effect relation between factor D (the amount of cholesterol in the diet) and factor X (the number of heart attacks); there is also a cause-and-effect relation between factor D and factor Y (the average blood cholesterol level). But most scientists would agree that it's an oversimplification to say this represents the whole picture. If you become a strict vegetarian and avoid cholesterol-containing foods altogether, there is no guarantee that you'll never have a heart attack. If you eat steak and eggs every day for breakfast, it doesn't mean that you are doomed to have a heart attack. The cause-and-effect relationship exists, but it's not "pure," and it's not absolute.

Note that when there is a cause-and-effect relationship between D and X, and also between D and Y, there can be a cause-and-effect relationship between X and Y, but that is not necessarily the case. Consider this example. Imagine X to represent the sentence "The visibility is poor on the highways." Let Y represent the sentence "People are using umbrellas." Let D represent "It is raining." Clearly, D can cause both X and Y. But X does not cause Y. Poor highway visibility does not, in itself, cause people to use umbrellas. (It might be foggy or snowing, but not raining.) Nor does Y cause X. If you get out your umbrella, it does not cause the highway visibility to become poor.

MULTIPLE FACTORS CAUSE BOTH *X* AND *Y*

If you watch television shows where the advertising is aimed at middle-aged and older folks, you'll hear all about cholesterol and heart disease. You might start wondering whether you should go to a chemistry lab to get your food. The cause-and-effect relationship between cholesterol and heart disease is complicated. The more we learn, it appears, the less we know.

Let's introduce and identify three new variables here: factor *S*, factor *H*, and factor *E*. Factor *S* is "Stress" (anxiety and frustration), factor *H* is "Heredity" (genetic background), and factor *E* is "Exercise" (physical activity). Over the past several decades, cause-and-effect relationships have been suggested between each of these factors and blood cholesterol levels, and between each of these factors and the frequency of heart attacks. Fig. 4-3 illustrates this sort of cause-and-effect relationship. Proving the validity of each link—for example, whether or not stress, all by itself, can influence cholesterol in the blood—is a task for future researchers. But every one of the links shown in the diagram has been suggested by somebody.

No matter how nearly the gray arrows in Figs. 4-2 and 4-3 represent true causation in the real world, they never approach the refined status of logical implication in a mathematical sense. Yet, all too often, that is what some people would like you to believe.

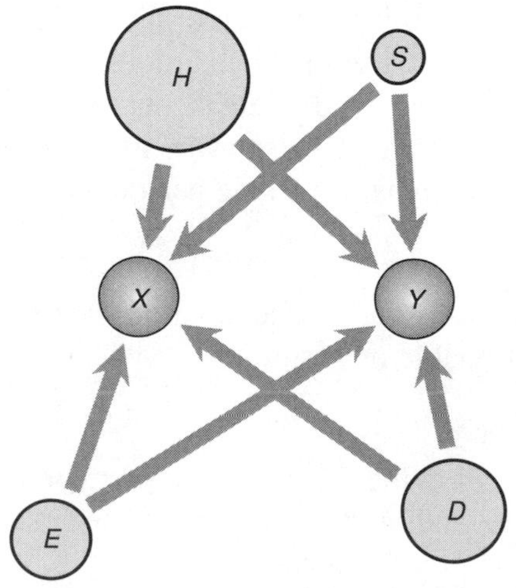

Fig. 4-3. A scenario in which there are multiple causes *D*, *S*, *H*, and *E* for both *X* and *Y*, not necessarily all of equal effect or importance.

THE PLAIN LOGIC

All of the preceding arguments and examples have been over one question that can be answered simply and finally by means of a truth table. Is the statement P & Q logically equivalent to the statement P ⇒ Q? Check it out for yourself. Compare the truth values for the two statements. The truth values do not match. The two statements are not logically equivalent.

PROBLEM 4-1

What are some reasonable cause-and-effect relationships that might exist in Fig. 4-3, other than those shown? Use single-shafted arrows to show cause and effect, and use the abbreviations we have used in the preceding discussion.

SOLUTION 4-1

Consider the following. Think about how you might conduct statistical experiments to check the validity of these notions, and to determine the extent of the correlation.

- H → S (Proposition: Some people are born more stress-prone than others.)
- H → D (Proposition: People of different genetic backgrounds have developed cultures in which the diets are different.)
- E → S (Proposition: Exercise can relieve or reduce stress.)
- E → D (Proposition: Extreme physical activity makes people eat more food, because they need more.)
- D → S (Proposition: Bad nutritional habits can worsen stress. Consider a hard-working person who lives entirely on coffee and potato chips, versus a hard-working person who follows a healthy diet.)

PROBLEM 4-2

What are some cause-and-effect relationships in the diagram of Fig. 4-3 that are questionable or absurd?

SOLUTION 4-2

Consider the following. Think about how you might conduct statistical experiments to find out whether or not the first three of these might be moved into the preceding category.

- H → E (Question: Do people of certain genetic backgrounds naturally get more physical exercise than people of other genetic backgrounds?)

- S → E (Question: Can stress motivate some types of people to exercise more, yet motivate others to exercise less?)
- S → D (Question: Do certain types of people eat more under stress, while others eat less?)
- S → H (Obviously absurd. Stress cannot affect a person's heredity!)
- E → H (Obviously absurd. Exercise cannot affect a person's heredity!)
- D → H (Obviously absurd. Dietary habits cannot affect a person's heredity!)

The Probability Fallacy

Here is a subtle but devastating fallacy that is so common that I have probably committed it at least a couple of times in this book already. We say something is true because we've seen or deduced it. If we believe something is true or has taken place but we aren't sure, it's tempting to say it is or was "likely." It's wise to resist this temptation.

BELIEF

When people formulate a theory, they often say that something "probably" happened in the distant past, or that something "might" exist somewhere, as-yet undiscovered, at this moment. Have you ever heard that there is a "good chance" that extraterrestrial life exists? Such a statement is meaningless. Either it exists, or it does not.

If you say, "I believe the universe began with an explosion," you are stating the fact that you believe it, not the fact that it is true or that it is "probably" true. If you declare, "The universe began with an explosion!" your statement is logically sound, but it is a statement of a theory, not a proven fact. If you say, "The universe probably started with an explosion," you are in effect suggesting that there were multiple pasts and the universe had an explosive origin in more than half of them. This is an instance of what can be called the *probability fallacy* (abbreviated PF), where probability is injected into a discussion inappropriately.

Whatever is, is. Whatever is not, is not. Whatever was, was. Whatever was not, was not. Either the universe started with an explosion, or else it did not. Either there is life on some other world, or else there is not.

PARALLEL WORLDS, FUZZY WORLDS

If we say that the "probability" of life existing elsewhere in the cosmos is 20%, we are in effect saying, "Out of n observed universes, where n is some large natural number, $0.2n$ universes have been found to have extraterrestrial life." That doesn't mean anything to those of us who have seen only one universe!

A word of defense ought to be inserted here concerning forecasts such as "The probability of measurable precipitation tomorrow is 20%," or "The probability is 50% that John Doe, a cancer patient, will stay alive for more than 12 months." These are based on past observations of large numbers of similar cases. A better way of stating the weather situation is "According to historical data, when weather conditions have been as they are today, measurable precipitation has occurred the next day in 20% of the cases." A better way of stating the cancer situation is "Of a large number of past cancer cases similar to that of John Doe, 50% of the patients lived beyond 12 months." These interpretations are understood. Therefore, these forecasters aren't guilty of committing the PF. Most people would take a dim view of a meteorologist who said, "Either it will rain tomorrow, or else it will not," or a doctor who said, "Either your dad will stay alive for more than 12 months, or else he will not."

It is worth mentioning that there are theories involving so-called *fuzzy truth*, in which some things "sort of happen." *Fuzzy logic* involves degrees of truth that range from completely false, through partially false, neutral, partially true, and totally true. Instead of only two values such as 0 (for falsity) and 1 (for truth), values can range along a continuum from 0 to 1. In some cases the continuum has other limits, such as 0 to 2, or −1 to +1. Entire books have been written about fuzzy logic, but we won't be getting into it.

WE MUST OBSERVE

Probability is usually defined according to the results of observations, although it is sometimes defined on the basis of theory alone. When the notion of probability is abused, seemingly sound reasoning can be employed to come to absurd conclusions. This sort of thing is done in industry every day, especially when the intent is to get you or me to do something that will cause somebody else to get rich. Keep your "probability fallacy radar" on when navigating through the real world.

If you come across an instance in which an author says that something "probably happened," "is probably true," "is likely to take place," or "is not likely to happen," think of it as another way of saying that the author believes or suspects that something did or didn't happen, is or isn't true, or is or is not expected take place on the basis of experimentation or observation.

Weak and Flawed Reasoning

When a rule of logic is broken in a mathematical system, any derived result in that system becomes suspect. It's possible to use flawed logic to come to a valid conclusion, but a flaw in reasoning often results in a mistaken conclusion. The trouble is especially great when the inaccurate conclusion has intuitive appeal—it "seems true"—and the error is not apparent unless or until a counterexample is found or its original proof is shown to be flawed.

"PROOF" BY EXAMPLE

A common fallacy is the use of specific examples to prove general statements. Sometimes we can get away with this, and sometimes we cannot. Consider the following statement:

- Some rational numbers are integers.

Let R represent the predicate "is a rational number," Z represent the predicate "is an integer," and x represent a variable from the set of numbers. Then the above statement can be symbolized as a wff, like this:

$$(\exists x)\ \mathrm{R}x\ \&\ \mathrm{Z}x$$

It's easy to prove that this is true. It's necessary only to show that it works for a single rational number, such as 35/5. That, of course, is equal to 7, and 7 is an integer. Once we've shown that the statement is true in one case, we've satisfied the existential quantifier "For some," which also means "There exists at least one."

But suppose instead we are confronted with this proposition:

- All integers are rational numbers.

When put into symbolic form, this proposition becomes a wff that contains a universal quantifier, like this:

$$(\forall x)\ \mathrm{Z}x \Rightarrow \mathrm{R}x$$

It is not difficult to find examples here. All we need to do is take an integer, such as 35, and then divide it by 1, getting 35 as the quotient, and then claim that the original integer is equal to this quotient. That's trivial:

$$35 = 35/1$$

This is a rational number, because it is a quotient in which the numerator is an integer and the denominator is a nonzero natural number. We can do this with many integers, always putting 1 in the denominator, generating all sorts of examples:

$$40 = 40/1$$
$$-45 = -45/1$$
$$260 = 260/1$$

Confident in our example-showing skills, we can arrange the set Z of all integers as a list, like this:

$$Z = \{0, 1, -1, 2, -2, 3, -3, \ldots\}$$

(Note that the set of integers is an uppercase italic Z, whereas the predicate "is an integer" is an uppercase non-italic Z. This differentiates between the two, and that is good, because a set differs from a predicate as much as a bus differs from the route it travels.) We can rewrite the set listing so every element is a quotient with 1 in every denominator, and therefore is obviously rational:

$$Z = \{0/1, -1/1, 1/1, 2/1, -2/1, 3/1, -3/1, \ldots\}$$

This proves the proposition for as many examples as we have the time and inclination to list: a hundred, a thousand, ten thousand, or a million. It strongly suggests that the proposition holds for all integers. Almost any reasonable person would come to the conclusion, after testing for a few specific integers, that the proposition is true. But merely "plugging in numbers" here does not prove the proposition and make it a theorem.

It is dangerous to use examples to prove general propositions that contain universal quantifiers. That requires more powerful tactics than the citing of examples. We can force the element in question to be a variable, such as x (and not a constant, such as 35) and use deductive logic, armed with the laws of arithmetic. Alternatively, we can apply mathematical induction to the set of integers after it has been arranged as a list with a defined starting point, as the set Z is portrayed above.

BEGGING THE QUESTION

You will sometimes hear or read "proofs" that do nothing more than assume the truth of the proposition to be proved. If you've a lawyer, you will recognize this fallacy right away when you see it. Have you ever pointed it out to your opponent in an argument or debate? Have you ever tried to get away with it in your own arguments? This is called *begging the question*

When you beg a question, you don't logically prove anything, whether the proposition is true or not; you prove only a triviality known as a *tautology*:

$$X \Rightarrow X$$

Here's an example of begging the question. Suppose the temperature is 40 degrees below zero Celsius, and the wind is gusting to 100 kilometers per hour (km/h). You state this fact and then conclude, "It is cold and windy today!" Here's another example. Suppose John Doe hit 100 home runs last baseball season. Your friend tells you this and then says, "John Doe hit a lot of home runs last year!" Neither of these are arguments. They're merely examples of the rephrasing or restatement of obvious truths.

Begging the question is often done in a more subtle manner. "It imperils the population to have motor vehicles moving at high speeds in residential areas. Therefore, if we allow people to drive cars and trucks on the streets of our cities at unlimited speeds, it presents a danger to the community." This merely says the same thing twice. It doesn't prove anything. We might as well say one thing or the other; it is pointless to say both.

Sometimes, begging the question takes a roundabout form in which there are several "logical steps," leading from the premise through a forest of logical maneuvers, and then back to the original premise. After that exercise, the fallacy-maker proudly proclaims, "*Q.E.D.*!"

HASTY GENERALIZATION

In the fallacy of *hasty generalization*, a certain characteristic is assigned to something as a whole, based on the examination of the wrong data, incomplete data, or data that is both wrong and incomplete.

Suppose that every time you ask people for favors when doing your laundry, they turn you down. What if this occurs a dozen times in a row? Let L represent the statement, "You are doing your laundry." Let F represent the statement, "You ask a person to do you a favor." Let T represent "The person does you the favor you just asked for." You find that (L & F) is repeatedly followed by ¬T. You use this experience to "prove" the following:

$$(L \& F) \Rightarrow \neg T$$

This is in effect an application of the fallacy of proof by example, applied to the real world. The fact that something has happened in numerous instances doesn't mean it happens in all instances, or even that it will happen in the very next instance.

CONTEXT

A word can have numerous meanings depending on the context in which it is used. But in a logical argument, we must not change the intended meaning of a word in the course of the discussion. This can result in absurd or nonsensical statements. "A pen that has run out of ink is no good for keeping cattle" is an example, where "pen" refers to a writing instrument first and an enclosure for animals second. That is an extreme example of what can happen with *improper use of context*.

CIRCUMSTANCE

Arguments are sometimes made in an effort to lead people to believe that a certain conclusion is reasonable. This is not true logic, because the conclusion is inferred but not proven. This is known as *argument by circumstantial evidence*.

Examples of arguments by circumstantial evidence are often heard in criminal trials. A lawyer "sets up" witnesses by asking questions not directly related to the crime. Imagine that I am accused of a crime. Some witnesses say they saw me in the vicinity of the place where the crime occurred; some witnesses testify to the effect that I was not home at the time of the crime; other witnesses express the opinion that I am a no-good, rotten son-of-a-buck. Even a thousand such testimonials do not *rigorously* imply that the I committed the crime. Even if my guilt can be inferred beyond "reasonable" doubt, it is not a *mathematical* proof of guilt.

SYLLOGISMS

A *syllogism* is an argument in which a conclusion is drawn based on two premises. The first premise is often a disjunction or an "if-then" statement. An example of a *disjunctive syllogism* is the following:

- Jill is in Florida or Jill is in New York. Jill is not in Florida. Therefore, Jill is in New York.

Let F represent the predicate "is in Florida," N represent "is in New York," and *j* represent the constant "Jill." Then this argument can be symbolized like this:

$$\begin{array}{l} \mathrm{F}j \vee \mathrm{N}j \\ \neg \mathrm{F}j \\ \hline \mathrm{N}j \end{array}$$

Here is an example of a syllogism containing a logical implication:

- Anyone who takes 100 sleeping pills all at once will die. Joe took 100 sleeping pills all at once. Therefore, Joe will die.

Let P represent the predicate "takes 100 sleeping pills all at once." Let D represent "will die." Let *x* be a logical variable, and let *j* represent the constant "Joe." Then symbolically, the argument looks like this:

$$\begin{array}{l} (\forall x)\ \mathrm{P}x \Rightarrow \mathrm{D}x \\ \mathrm{P}j \\ \hline \mathrm{Dj} \end{array}$$

We can refute (disagree with or disprove) one or the other of the premises in either of these syllogism examples, but in themselves, the arguments are logically valid.

A common fallacy in syllogisms is that of *denying the antecedent*. An example is the following argument:

- If you commit a federal offense, you'll go to prison. You did not commit a federal offense. Therefore you will not go to prison.

Let F represent the predicate "commit(s) a federal offense." Let P represent "will go to prison." Let *y* represent the constant "you." Then the above argument looks like this in symbolic form:

$$\begin{array}{l} \mathrm{F}y \Rightarrow \mathrm{P}y \\ \neg \mathrm{Fy} \\ \hline \neg \mathrm{P}y \end{array}$$

This is not a logically valid argument. There are plenty of non-federal crimes that can land you in prison if you commit them. Some innocent people, who commit no crimes at all, also end up in prison!

The foregoing fallacy can also occur if the original antecedent is negative:

- If John was not near the grocery store last night, he must have been at home. John was near the store last night. Therefore, he couldn't have been at home.

Let G represent the predicate "was near the grocery store last night." Let H represent "was at home." Let *j* represent the constant "John." Then symbolically:

$$\begin{array}{l} \neg Gj \Rightarrow Hj \\ Gj \\ \hline \neg Hj \end{array}$$

This reasoning is not valid! What if John's home is next to the grocery store? Then he is near the grocery store even when he is at home. In addition to that flaw, this argument contains a lack of clarity in the meaning of the word "near." Lawyers will recognize this sort of trick!

Another fallacy can occur in disjunctive syllogisms. Consider the following dilemma and argument:

- Wanda must leave the country or get arrested for a crime of which she has been accused. Wanda has left the country. Therefore, Wanda will not get arrested.

Let L represent the predicate "must leave the country." Let A represent "will get arrested for a crime of which she has been accused." Let *w* represent the constant "Wanda." Then symbolically, our argument looks like this:

$$\begin{array}{l} Lw \vee Aw \\ Lw \\ \hline \neg Aw \end{array}$$

This argument is fallacious. Wanda might get arrested even if she leaves the country. This fallacy arises from confusion between the inclusive and exclusive forms of the operation "or."

FUN WITH SILLINESS

It can be fun to use nonsensical subject-predicate combinations to show logical validity or invalidity. This short-circuits the human tendency to assign notions about everyday life to logical derivations. Consider this:

- If the moon is made of Swiss cheese, then some ants eat chocolate. The moon is made of Swiss cheese. Therefore, some ants eat chocolate.

Let S represent the predicate "is made of Swiss cheese." Let A stand for "is an ant." Let C stand for the predicate "eats chocolate." Let *m* represent the constant "the moon," and let *x* represent a logical variable. Then the above argument is symbolized like this:

$$\mathrm{S}m \Rightarrow [(\exists x)\ \mathrm{A}x \ \&\ \mathrm{C}x]$$
$$\mathrm{S}m$$

$$(\exists x)\ \mathrm{A}x \ \&\ \mathrm{C}x$$

The argument is logically valid. But suppose we deny the antecedent. Then we can argue that because the moon is not made of Swiss cheese, no ant will eat chocolate. That is fallacious. I'll bet hungry ants would swarm all over a warm chocolate bar if it were made available to them, regardless of the material composition of the moon.

Now scrutinize this:

- Either Mars is inhabited by little green rabbits, or the sky appears blue as seen from the surface of the earth on a clear day. Mars has no little green rabbits. Therefore, the sky is blue as seen from the surface of the earth on a clear day.

Let R represent "is inhabited by little green rabbits." Let B represent "appears blue as seen from the surface of the earth on a clear day." Let *m* represent the constant "Mars." Let *s* represent the constant "the sky." Then the above argument looks like this when symbolized:

$$\mathrm{R}m \vee \mathrm{B}s$$
$$\neg\mathrm{R}m$$

$$\mathrm{B}s$$

This illustrates a situation where the first statement (in this case, a disjunction) in a syllogism is always true. The logic is valid, but the example proves nothing. If there were little green rabbits on Mars, the sky would nevertheless be blue as seen from the surface of the earth on a clear day. By plugging in other predicates and constants, logic tricksters often convince people that two events are connected, when in fact they are not.

INDUCTIVE REASONING

In *inductive reasoning*, we attempt to show that something is true most of the time, or that it is reasonable to expect that it will occur. This is not the same thing as mathematical induction. Inductive reasoning is often presented as rigorous deduction, fooling people into thinking that arguments are air-tight when they are really full of holes. To make things worse, a conclusion is often stated to the effect that something "is probably true" or "will probably occur" or "probably took place," thereby invoking PF in addition to the use of flimsy reasoning masquerading as mathematical logic.

Here is an example of what can happen when inductive reasoning is combined with the PF, generating an absurd conclusion. Suppose the speed limit on a stretch of highway is 100 kilometers per hour (km/h). An officer who needs to meet a quota interprets this to mean that you are speeding if you drive at *100 km/h or more* (as opposed to *more than 100 km/h*). Imagine you are cruising along at 99.6 km/h and the police radar reads the speed digitally to the nearest kilometer per hour, rounding it off to 100 km/h. The officer sees this and concludes that you are "probably" speeding. His reasoning goes as follows. Given the radar reading, the probability that you are going 100 km/h or more is exactly 50%, because your exact, true speed must be more than 99.5 km/h but less than 100.5 km/h. If we round off 50%, or 0.5, to the nearest whole digit, then that digit is by convention equal to 1, or 100%. This means that a probability of 0.5 is equivalent to a probability of 1! So in the officer's fallacy-tormented mind, a reading of 100 km/h on radar means you are going 100 km/h or more if you drive 99.5 km/h or more!

Bring on a good lawyer with a mind for mathematical logic, a few thousand dollars, a whimsical judge, and an intelligent jury, and I will pay money to watch the trial of this case!

PROBLEM 4-3

Suppose someone makes the following statement and claims that it is a mathematical theorem:

- All rational numbers can be written as terminating decimals, that is, as decimal numbers where the digits after a certain point are all zeroes.

You claim that this cannot be a theorem because it is not true. How many counterexamples must you find in order to show that this theorem is not true?

SOLUTION 4-3

You need to find only one counterexample to show that a claimed theorem is not true. In this case, an example is $\frac{1}{3} = 0.333\ldots$, which is not a terminating decimal because the numeral 3 keeps repeating without end.

PROBLEM 4-4

What sort of fallacy is committed in the following argument? Symbolize it, and then identify the fallacy. Note that a *polygon* is a geometric figure that lies entirely in a single plane, and that has three or more straight sides such that all adjacent pairs of sides intersect at their end points, and no two sides intersect except at their end points. A *triangle* is a polygon with three sides.

- All triangles are polygons. Figure *S* is not a triangle. Therefore, figure *S* is not a polygon.

SOLUTION 4-4

You know, of course, that there are plenty of polygons besides triangles. Squares are good examples. So are rectangles, trapezoids, pentagons, and infinitely many other types of figures. The above argument is obviously flawed, but how? Let's symbolize it. Let T represent the predicate "is a triangle," let P represent the predicate "is a polygon," let x represent a logical variable, and let s represent the constant "figure *S*." The above argument can then be rearranged like this:

- For all x, if x is a triangle, then x is a polygon. It is not true that s is a triangle. Therefore, it is not true that s is a polygon.

Symbolically:

$(\forall x)\ \mathrm{T}x \Rightarrow \mathrm{P}x$
$\neg \mathrm{T}s$

$\neg \mathrm{P}s$

This is an example of denying the antecedent. Note that the argument can be turned around and it becomes sound. Symbolically:

$$(\forall x)\ \mathrm{T}x \Rightarrow \mathrm{P}x$$
$$\neg \mathrm{P}s$$
$$\overline{\qquad\qquad\qquad}$$
$$\neg \mathrm{T}s$$

This translates to:

- All triangles are polygons. Figure *S* is not a polygon. Therefore, figure *S* is not a triangle.

Paradoxes and Brain Teasers

Here are some tidbits that seem to defy logic. These sorts of things can be fun or frustrating, depending on your disposition. In any case, they show what can happen when attempts are made to apply mathematical rigor to decidedly non-rigorous realities and ideas.

THE MEANING OF RANDOMNESS

Generating a sequence of *random numbers* seems like an easy job at first. All we need to do is rattle off digits from 0 to 9 in any ridiculous sequence we please, and the result will be a string of "random" numbers—right? Wrong! Any person, if tested, will show a leaning or preference for certain digits or sequences of digits, such as 5 or 58. People are biased, even when it comes to something as generic as their taste in numbers! A truly random sequence of numbers will have no bias whatsoever.

Another characteristic of randomness is unpredictability. There should be no way of generating the next digit in a sequence of random digits, based on the previous ones. If there is, the sequence can't be random, because that next digit is predetermined. We can get digits for the decimal expansion of the value of the square root of 2, or any other positive integer, by a using a scheme called *extraction of the square root*. If we have the patience, and if we know the first *n* digits of the decimal expansion of a square root, we can find, or "predict," the $n + 1$st digit by means of this process. That disproves the notion that the square root of 2, or of any other positive integer, has a decimal expansion with truly random digits. There are also processes for finding, or "predicting," the digits in the decimal expansions of other types of irrational numbers such as pi (π), which is the ratio of the circumference of any circle or sphere to its diameter.

If no one can chatter off random digits, and if the digits in irrational numbers never occur in random sequence, where can we find digits that do occur in truly random fashion? Maybe we can't!

Suppose it is impossible for any human being or machine to ever find a truly random sequence of numbers if we require that the digits be unpredictable! This is a haunting, and quite plausible, idea. If truly random digits are unpredictable, then they cannot be generated by any definable process, and this includes the thoughts of the most jumbled-up human brain, or the digital operations of the most exotically programmed computer. If there is no way to generate a sequence of truly random digits, then even if such sequences exist, they are beyond our ability to observe.

Fortunately, in applications that require the use of random digits, so-called *pseudorandom numbers* are usually good enough. A good way to get a pseudorandom sequence of digits is to use an electronic meter to periodically measure the intensity of radio-frequency noise, such as the "hiss" generated by the movement of electrons among atoms, or the emissions from far-off galaxies in outer space. A discussion of those processes is beyond the scope of this book!

A WIRE AROUND THE EARTH

We are all familiar with the irrational number π, which is equal to approximately 3.1416, and represents the number of diameters in the circumference of a circle or sphere. This number has been known for thousands of years to be a constant that does not depend on the size of the circle or sphere.

An interesting counterintuitive result can be derived from simple application of the formula for the circumference of a circle or sphere to its diameter. The formula is:

$$c = \pi d = 2\pi r$$

where c is the circumference, d is the diameter, and r is the radius, all in the same units. Suppose the earth were a smooth, perfectly round sphere, with no hills or mountains. Imagine a perfectly non-elastic wire around the equator, strung so that it is snug and does not stretch. If we add 10 meters (10 m) to the length of this wire, and then prop it up all the way around the planet so that it stands out equally far everywhere, how far above the surface will it stand? Assume the circumference of the earth is 40,000,000 meters.

Most people are inclined to think that the wire will stand out only a tiny distance from the surface of the earth if 10 meters is added to its length. After all, that is an increase of only 10 parts in 40,000,000, or 0.000025%. But in fact the wire will stand out approximately 1.59 meters all the way around the sphere.

PROBLEM 4-5

We don't have to be content with showing how this works for the earth only. The assertion can be extended to claim that when 10 meters is added to the length of a wire that tightly girdles the circumference of any sphere, no matter how big or small, the lengthened wire will stand out the same distance from the surface: approximately 1.59 meters. Prove this!

SOLUTION 4-5

Refer to Fig. 4-4. Suppose the radius of the sphere, expressed in meters, is equal to r. Suppose the circumference of the sphere, also expressed in meters, is equal to c. From the rules of Euclidean spatial geometry, we know the following:

$$c = 2\pi r$$

Solving the equation for r gives us this:

$$r = c/(2\pi)$$

If we add 10 meters to the length of a wire whose original length is equal to c (because it girdles the sphere around a circumference), then

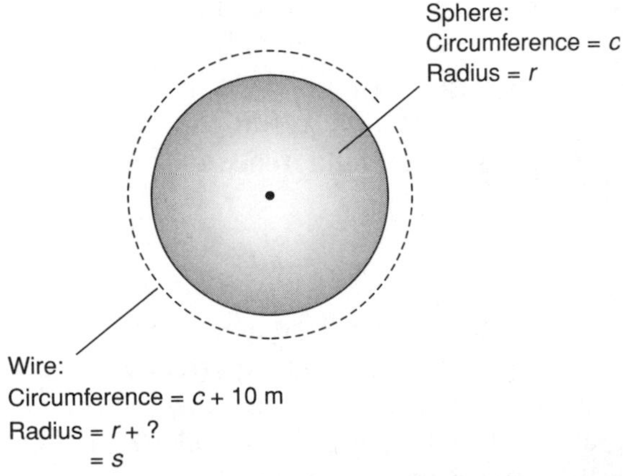

Fig. 4-4. If the circumference of a wire around a planet is increased by 10 m, by how much does the radius increase?

the lengthened wire lies in a circle whose radius s, expressed in meters, can be calculated as follows if we let $\pi = 3.14$:

$$
\begin{aligned}
s &= (c + 10)/(2\pi) = c/(2\pi) + 10/(2\pi) \\
&= c/(2\pi) + 1.59 \\
&= r + 1.59
\end{aligned}
$$

This shows that the radius of the circle described by the lengthened wire is 1.59 meters greater than the radius of the sphere. Therefore, the lengthened wire can be positioned so that it stands 1.59 meters above the surface of the sphere, all the way around.

THIS STATEMENT IS FALSE

A common logical paradox is often cited in the self-contradicting assertion "This statement is false." If we affirm the statement, then it is true, and this contradicts its assertion that it is false. If we deny the statement, then we assume it is false, but it's not true that it's false, so therefore it's true. A statement such as this is meaningless, because it cannot be true and cannot be false.

Russell's Paradox is a more sophisticated example. We can state it as follows. Suppose Hap is a barber in the town of Happyton, and he shaves all the people, but only the people, in Happyton who don't shave themselves. Does Hap shave himself or not?

Assume Hap shaves himself. We've just stated that he shaves only those people who do not shave themselves! Therefore, Hap does not shave himself. This means Hap shaves himself and Hap does not shave himself, and that is a contradiction. By *reductio ad absurdum*, Hap does not shave himself. From this, it follows that Hap shaves himself, because he shaves all the people (including himself) that do not shave themselves. Again, *reductio ad absurdum* can be invoked, and we find that Hap shaves himself. We have set ourselves running in a circle of contradictions, like a dog chasing its own tail when it doesn't even have a tail!

The only way out of this quagmire is to conclude that there can be no such person as Hap.

THE FROG AND THE WALL

A familiar problem in mathematics is the adding-up, or *summing*, of an *infinite sequence* in order to get a finite sum. The frog-and-wall paradox (or what at first seems to be a paradox) shows an example of this.

Suppose there is a frog at a certain distance from a wall, say, 8 meters. Imagine that this frog jumps halfway to the wall, so that she is 4 meters away. Suppose she continues to jump toward the wall, each time getting halfway there. She will never reach the wall if she jumps in this way, no matter how many times she jumps, even though she has only 8 meters to travel at the outset. The frog will die long before she gets to the wall, even though she almost reaches it (Fig. 4-5). No finite number of jumps will allow the frog to reach the wall. That would take an infinite number of jumps.

This scenario can be based on the following *infinite series*. (A *series* is the sum of the terms in a *sequence*.) Let's call it S:

$$S = 4 + 2 + 1 + \frac{1}{2} + \frac{1}{4} + \frac{1}{8} + \ldots$$

If we keep cutting a number in half, over and over, and add the result, the final sum of this type of infinite series is twice the original number. That means $S = 8$. But how is it possible to add up an infinite number of numbers? In the real world, of course, it isn't. We can only approach the actual sum in real life, because there isn't enough time (that is, an infinite amount of time) to add up an infinite number of numbers. But in the mathematical world, certain infinite series add up to finite numbers. Such a series is said to be *convergent*.

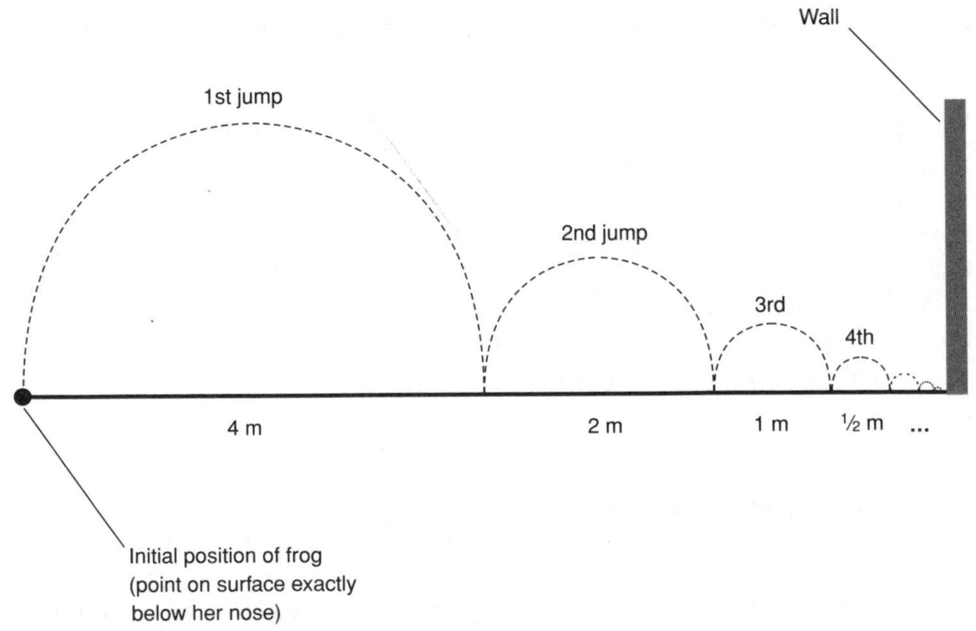

Fig. 4-5. A frog jumps towards a wall, getting halfway there each time.

A real-world frog cannot reach the wall by jumping halfway to it, over and over. But a theoretical one can! There are two ways this can happen. First, in the mathematical world, there is an infinite amount of time, so an infinite number of jumps can take place. Another way around the problem is to keep halving the length of time in between jumps, say from four seconds to two seconds, then to one second, then to half a second, and so on. This will make it possible for the frog to hop an infinite number of times in only a finite span of time!

PROBLEM 4-6

The same sort of argument as the one in the frog-and-wall scenario can be used to "prove" that if you are driving at 80 km/h and trying to pass someone ahead of you who is going 50 km/h, you will never catch, let alone pass, that driver. How is this "proof" done? What's wrong with it?

SOLUTION 4-6

Fig. 4-6 is a geometric illustration of this situation. The initial state of affairs is shown at A; you, going 80 km/h, are a certain distance d_0 behind the car you are trying to catch, which is going 50 km/h. After a certain time you have traveled the distance d_0, and are in the position previously occupied by the other car. But that car has moved ahead by a distance d_1, so it is in a new position. The distance d_1 is less than the distance d_0. The situation after you have traveled the distance d_0 is shown in Fig. 4-6B.

Once you have traveled the distance d_1, the other car has moved ahead and is in front of you by a distance d_2, as shown in Fig. 4-6C. Figs. 4-6D and E show what happens in the next two time frames, as you travel distance d_2 and then distance d_3. At E, the car is ahead of you by distance d_4. This process goes on without end. Therefore, you can never catch the other car!

In real life, you will catch and pass the slower driver because the sequence $d_0, d_1, d_2, d_3, d_4, \ldots$ is such that its corresponding series, T, is convergent:

$$T = d_0 + d_1 + d_2 + d_3 + d_4 \ldots$$

The trick here lies in the fact that the sum of the time intervals corresponding to the transitions of the distance intervals is finite, not infinite. You span the distances $d_0, d_1, d_2, d_3, d_4, \ldots$ at an ever-increasing rate when you race down the highway to pass that slowpoke in front of you. The rate at which you span the progressively smaller intervals

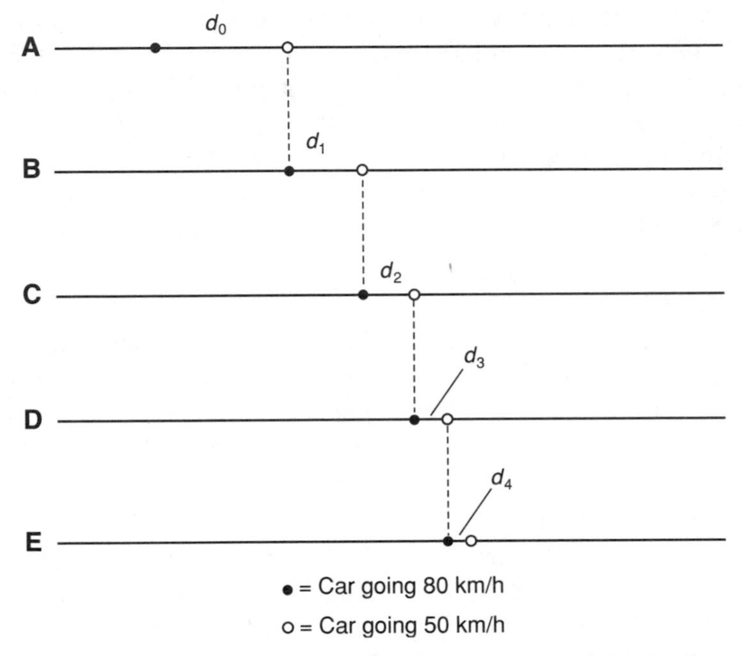

Fig. 4-6. Illustration for Problem 4-6. A fast car comes up behind a slower car. The distances d_0, d_1, d_2, d_3, d_4,...keep getting smaller.

"blows up." You add up an infinite number of numbers, corresponding to smaller and smaller distance intervals, in a finite length of time. This is the same sort of thing that happens in the frog-and-the-wall scenario.

A GEOMETRY TRICK

Here is a geometry puzzle that, at first glance, seems to defy the laws of Euclidean geometry. It is an example of the misuse of drawings to help a would-be deceiver come to an invalid conclusion. Fig. 4-7 illustrates the scheme, and its resolution, in four stages.

In Fig. 4-7A, a square is divided into 64 square units, 8 on a side, and then is cut along the indicated lines, making two right triangles, **X** and **Z**, measuring 3-by-8 units, and two trapezoids, **W** and **Y**, consisting of 3-by-6 rectangles added to 2-by-5 right triangles. These four pieces are rearranged to create a 13-by-5 rectangle (Fig. 4-7B). The area of the rectangle is 13 × 5, or 65 square units.

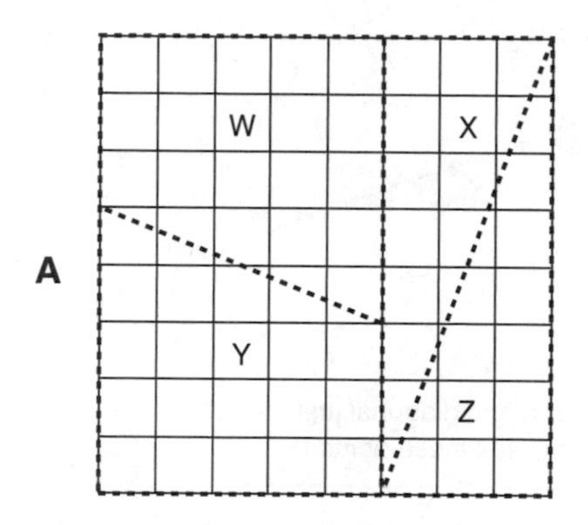

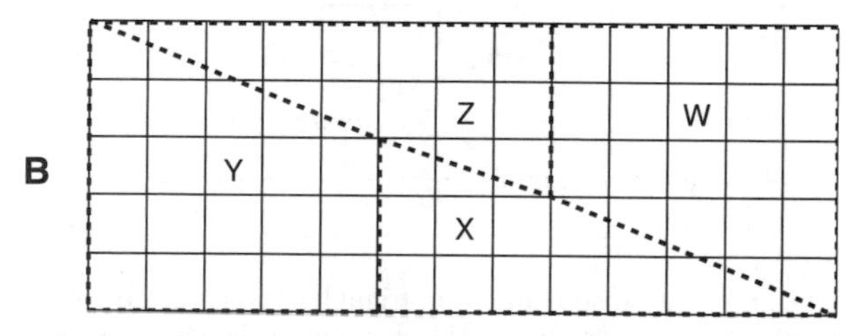

Fig. 4-7A,B. At A, a square is divided into four sections, called W, X, Y, and Z. At B, the four sections are reassembled into a rectangle with an area that appears to be 1 square unit larger than that of the large square. Illustrations for Problem 4-7.

By merely cutting up and rearranging the square, we've created 1 square unit of area from nothing!

PROBLEM 4-7

Something must be wrong with the rearrangement process shown in Fig. 4-7. Where does the "phantom area" come from?

SOLUTION 4-7

The trouble is in the center of the rectangle shown in Fig. 4-7B. There is a rectangle there, as shown in Fig. 4-7C, that measures 1-by-3 units. This rectangle can be sliced into two 1-by-3 right triangles, as shown. Just to the left and right of this central rectangle, there are two 2-by-5 right triangles. If the long diagonal through the 13-by-5 rectangle were a straight line, then the ratios 1:3 and 2:5 would be equal. But the ratios are not equal! The long diagonal cutting the 13-by-5 rectangle in Fig. 4-7C is not a straight line. If it were, we would have the scenario of Fig. 4-7D, where the long diagonal line ramping down the 13-by-5

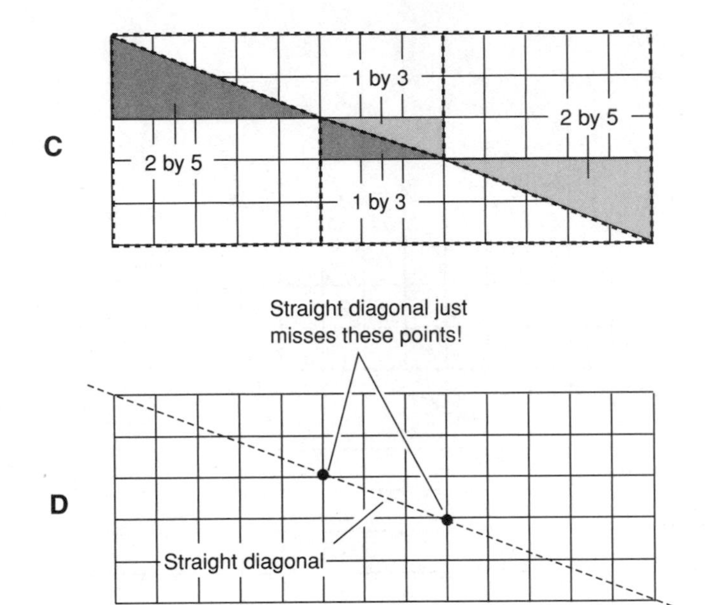

Fig. 4-7C,D. At C, illustration showing right triangles "constructed" within the rectangle. At D, an actual straight diagonal of the large rectangle doesn't intersect any of the vertices of the small squares. Illustrations for Problem 4-7.

rectangle would not intersect any of the vertices of any of the small squares. The visual deception occurs over such an elongated shape that it is difficult to see in drawings.

A "PROOF" THAT −1 = 1

There are several ways to "prove" that −1 = 1. All such "proofs" involve subtle reasoning flaws, or the neglect of certain facts of arithmetic. Table 4-1 shows one method of "proving" that −1 = 1. This table is an example of a *statements/reasons (S/R) proof*, a popular way to abbreviate demonstrations that proceed in neat steps. In this "proof," the square root of a quantity is denoted as the quantity raised to the ½ power. That is, the ½ power means exactly the same thing as a radical sign for the purposes of this discussion. We start with an obviously true statement, and from it, apparently using the rules of arithmetic, we derive something that is obviously false.

PROBLEM 4-8

What is wrong with the "proof" portrayed in Table 4-1?

SOLUTION 4-8

Perhaps you wonder whether we have any business talking about the square root of a negative number. That's not the problem here; the square roots of negative numbers are defined. The problem with the "proof" shown in Table 4-1 lies in the fact that the square root of any quantity can be positive or negative. We normally think the square root of 1 is equal to 1. But it can also be equal to −1. If we multiply either 1 or −1 by itself, we get 1. In a sense, then, the square root of 1 has two different values! The trick in the "proof" shown in Table 4-1 lies in the exploitation of this fact. We subtly take advantage of the duality, and conclude by saying that one part of the "two-valued" square root of 1 is equal to the other part.

Theories have been developed around the notion of multi-valued numbers. These theories produce a lot of interesting results, and have applications in physics and engineering. Theories of multi-valued numbers also get rid of apparent paradoxes and contradictions like the one here.

Table 4-1. A "proof" that −1 = 1. The ½ power of a quantity denotes the square root of that quantity. Each line in the table proceeds from the previous line, based on the reason given.

Statements	Reasons
$(-1)^{1/2} = (-1)^{1/2}$	A quantity is always equal to itself.
$[1/(-1)]^{1/2} = [(-1)/1]^{1/2}$	Both $1/(-1)$ and $(-1)/1$ are equal to −1. Either of these expressions can be substituted for −1 in the previous equation.
$1^{1/2}/(-1)^{1/2} = (-1)^{1/2}/1^{1/2}$	The square root of a quotient is equal to the square root of the numerator divided by the square root of the denominator.
$(1^{1/2})(1^{1/2}) = [(-1)^{1/2}][(-1)^{1/2}]$	Any pair of equal quotients can be cross-multiplied. The numerator of the one times the denominator of the other equals the denominator of the one times the numerator of the other.
$(1^{1/2})^2 = [(-1)^{1/2}]^2$	Either side of the previous equation consists of a quantity multiplied by itself. That is the same thing as the quantity squared.
$1 = -1$	When the square root of a number is squared, the result is the original number. Therefore, all the exponents can be taken out of the preceding equation.

THE WHEEL PARADOX

Here's a famous paradox that concerns a pair of concentric wheels. Both are rigidly attached to each other so that one rotation of the large wheel is attended by exactly one rotation of the smaller wheel (Fig. 4-8).

Imagine that this wheel rolls along a double surface, as shown in the diagram, so that the large wheel makes exactly one rotation along the lower surface. If the diameter of the larger wheel is d_1, then the length of the path along the surface for one rotation is πd_1.

The upper surface is spaced just right with respect to the lower surface, so the smaller wheel can move along the upper surface while the larger wheel moves along the lower surface. The smaller wheel has a diameter of d_2. The smaller wheel rotates at the same rate as the larger wheel, because the two are rigidly attached. When the larger wheel rotates 360°, the smaller wheel does, too. But the smaller wheel traverses the same distance (πd_1) as the larger wheel. This implies that the two wheels must have the same circumference, even though their diameters are different.

PROBLEM 4-9

How is the previous paradox resolved? Two wheels with different diameters cannot have the same circumference, can they?

SOLUTION 4-9

The catch lies in the fact that the smaller wheel slides, or skids, along the upper surface as the larger wheel rolls along its surface with good traction. Nothing in the statement of the problem forbids skidding!

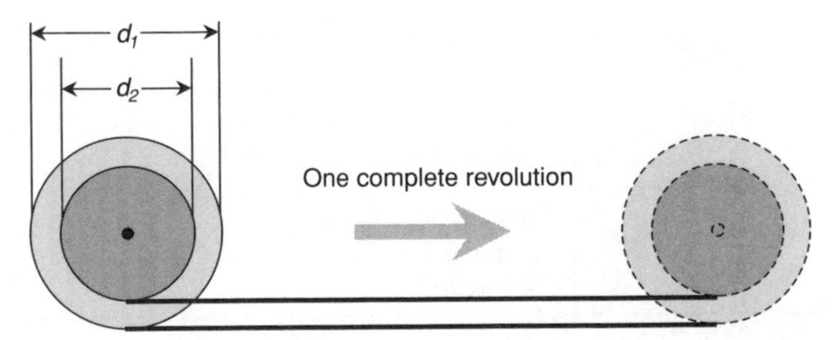

Fig. 4-8. The wheel paradox. The larger wheel has diameter d_1, and the smaller wheel has diameter d_2. Illustration for Problem 4-9.

Quiz

This is an "open book" quiz. You may refer to the text in this chapter. A good score is 8 correct. Answers are in the back of the book.

1. Suppose you want to prove the proposition $(\exists x)$ P$x \vee$ Qx. Let k be a constant, and an element of the set for which the variable x is defined. In order to prove the proposition using the constant k, the minimum that we must do is show the truth of the statement
   (a) P$k \vee$ Qk.
   (b) Pk.
   (c) Qk.
   (d) Any of the above

2. Let x be a logical variable. Let A stand for the predicate "is an apple." Let P stand for the predicate "is purple." Suppose you are confronted with the following proposition:

$$(\exists x)\ \mathrm{A}x\ \&\ \mathrm{P}x$$

   In order to disprove this proposition, you must prove that
   (a) If x is not an apple, then x is probably purple.
   (b) If x is not purple, then x is not an apple.
   (c) If x is an apple, then x is not purple.
   (d) Any of the propositions (a), (b), or (c) above

3. Suppose a strong correlation is found between the wobbulation of widgets and the diddlefaction of doodads in the town of Warpington during the Year of the Fiddle. This logically and rigorously implies
   (a) that the relationship between widget wobbulation and doodad diddlefaction in Warpington during the Year of the Fiddle was a coincidence; there was no causation of any sort involved.
   (b) that widget wobbulation caused doodads to diddlefact in Warpington during the Year of the Fiddle.
   (c) that doodad diddlefaction caused widgets to wobbulate in Warpington during the Year of the Fiddle.
   (d) None of the above

4. Suppose you invent a new mathematical operation and call it "retrofaction." You symbolize it by means of an uppercase Greek letter theta (Θ). Suppose you discover, using a computer programmed to perform the operation of retrofaction, that the following holds true for all the positive integers n and p from 1 up to 1,000,000,000:

$$(\forall n, p)\ (n \ \Theta\ p = p \ \Theta\ n)$$

What does this prove?

(a) If x and y are real numbers, then:

$$(\forall x, y)\ (x \ \Theta\ y = y \ \Theta\ x)$$

(b) If x and y are rational numbers, then:

$$(\forall x, y)\ (x \ \Theta\ y = y \ \Theta\ x)$$

(c) If x and y are positive integers, then:

$$(\forall x, y)\ (x \ \Theta\ y = y \ \Theta\ x)$$

(d) None of the above

5. If a lawyer demonstrates that a defendant is guilty beyond a reasonable doubt, that lawyer makes use of

(a) denial of the antecedent.
(b) *reductio ad absurdum.*
(c) inductive reasoning.
(d) DeMorgan's laws.

6. Suppose the weather service comes over a radio station with this announcement: "A thunderstorm has just passed over Stonyburg. Several observers reported seeing baseball-sized hail stones in Stonyburg within the last hour. Therefore, this is a hail-producing thunderstorm." This is an example of

(a) the PF.
(b) begging the question.
(c) the use of a word in improper context.
(d) an attempt to prove a generality by example.

7. According to some portrayals, the Roman emperor Caligula had blond hair on the day he declared himself to be one of the gods. How likely or possible is it that this was in fact the case?

(a) Unlikely
(b) Quite likely
(c) Almost certain
(d) None of the above

8. Suppose that an employee lets a building burn down because of negligence. The owner of the building sues him. The plaintiff's lawyer argues that such an employee must be a totally irresponsible sort of person, and that a responsible person would have been able to prevent the fire or put it out before it spread. The defense concedes this, especially in light of the fact that several psychiatrists have testified that the defendant has no sense of responsibility whatsoever. Then the defense attorney goes on to say that because of psychological deficiency on the part of the defendant, the defendant cannot be responsible for anything, even the burning-down of a building because of his own negligence. This argument is
   (a) perfectly sound.
   (b) fallacious, because it tries to apply the same meaning to a word in two different contexts.
   (c) fallacious, because it commits the PF, and probability has nothing to do with this situation.
   (d) fallacious, because it attempts to prove a generality by example, and a generality can never be proved by example.

9. In order to disprove a theorem that says something is true for all elements of a set S, it is necessary to
   (a) find only one element in S for which the theorem does not hold true.
   (b) find two or more elements in S for which the theorem does not hold true.
   (c) show that the theorem holds true for less than half the elements of S.
   (d) prove another theorem that states exactly the opposite for all elements of S.

10. Suppose you want to prove the proposition $(\exists y)\ \mathrm{M}y\ \&\ \mathrm{N}y$. Let k be a constant, and an element of the set for which the variable y is defined. In order to prove the proposition, the minimum that we must do is show the truth of the statement
   (a) $\mathrm{M}k\ \&\ \mathrm{N}k$.
   (b) $\mathrm{M}k$.
   (c) $\mathrm{N}k$.
   (d) Any of the above

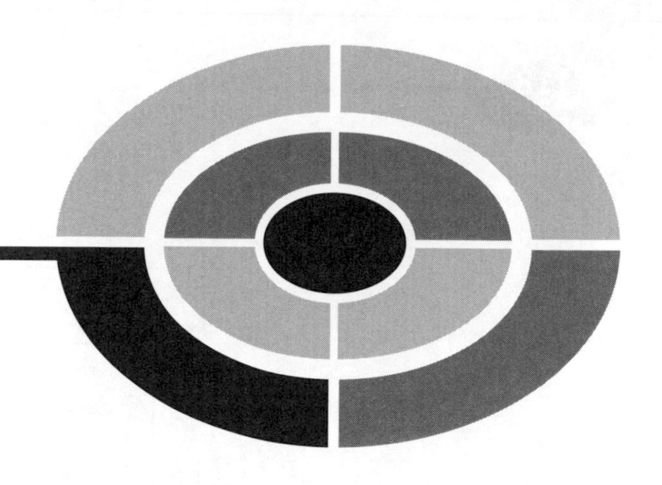

Test: Part One

Do not refer to the text when taking this test. You may draw diagrams or use a calculator if necessary. A good score is at least 30 answers (75% or more) correct. Answers are in the back of the book. It's best to have a friend check your score the first time, so you won't memorize the answers if you want to take the test again.

1. The conjunction of seven sentences is true
   (a) if and only if all the sentences are true.
   (b) if and only if at least one of the sentences is true.
   (c) if and only if at least two of the sentences are true.
   (d) if and only if all the sentences are false.
   (e) under no circumstances, because a conjunction can't be defined for more than two sentences.

2. Examine Fig. Test 1-1. This diagram applies to two types of objects, known as doodads and widgets. Let D symbolize the predicate "is a doodad," and let W symbolize the predicate "is a widget." Let x be a logical variable. Let f, g, and h be constants. Which of the following statements is true?

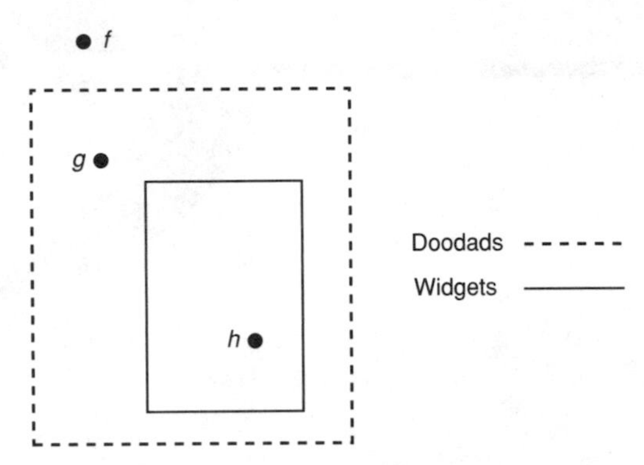

Fig. Test 1-1. Illustration for Part One Test Questions 2 and 3.

(a) $(\forall x)$ D$x \Rightarrow$ Wx
(b) $(\forall x)$ W$x \Rightarrow$ Dx
(c) $(\forall x)$ $\neg$D$x \Leftrightarrow$ Wx
(d) $(\forall x)$ D$x \Rightarrow \neg$Wx
(e) $(\forall x)$ $\neg$D$x \Rightarrow$ Wx

3. Examine Fig. Test 1-1. This diagram applies to two types of objects, known as doodads and widgets. Let D symbolize the predicate "is a doodad," and let W symbolize the predicate "is a widget." Let x be a logical variable. Let f, g, and h be constants. Which, if any, of the following statements (a), (b), (c), or (d) is false?
   (a) Dh & Wh
   (b) D$h \vee$ Wh
   (c) $\neg$Df & $\neg$Wf
   (d) D$g \vee$ Wg
   (e) All of the above statements (a), (b), (c), and (d) are true.

4. A lemma is
   (a) a definition used to prove another definition.
   (b) a theorem that results from implication reversal.
   (c) a corollary to an existing theorem.
   (d) a theorem used as a shortcut in the proof of another theorem.
   (e) a non-rigorous axiom.

5. In logic, the expression "X implies Y" means that
   (a) If X is true, then Y is true.
   (b) If X is true, then Y is probably true.

(c) If X is true, then there is good reason to believe that Y is true.
(d) If X is true, it suggests that Y is true, but there are exceptions.
(e) If X is true, then Y is true; and if Y is true, then X is true.

6. Consider the statement "I am a swimmer." This sentence contains
   (a) a subject, a verb, and an object.
   (b) a subject and a verb only.
   (c) a subject, a linking verb, and a complement.
   (d) a subject and a linking verb only.
   (e) a subject, a verb, and a linking verb.

7. If a contradiction is found in a mathematical system, then
   (a) all previously proven theorems in the system become false.
   (b) you can't be sure that any theorem in the system is true.
   (c) all the axioms become invalid.
   (d) all the definitions become meaningless.
   (e) nothing is wrong; contradictions are perfectly all right.

8. In the sentence "You run to the library," the word "library" is
   (a) the subject.
   (b) the verb.
   (c) the complement.
   (d) the object.
   (e) the predicate.

9. Suppose that you are building up a mathematical theory. You are able to prove some proposition Q. Later on in the theory, you find that you have proven $\neg Q$. What does this indicate?
   (a) Everything is all right, and you should create some more axioms.
   (b) The theory is too strong, so you should get rid of some definitions to make it weaker.
   (c) Your set of theorems is consistent, and you have made no mistakes of any kind.
   (d) You haven't proved enough theorems, so you need to prove some more.
   (e) You have made a mistake in one of your proofs, or the theory is flawed.

10. Mathematical induction is a technique that makes it possible to
    (a) prove a proposition for an infinite number of elements, using a finite number of steps.
    (b) prove a proposition by deriving a contradiction from its negation.

(c) prove a proposition by implication reversal.
(d) prove a proposition directly from a single axiom.
(e) prove that a proposition is probably true.

11. In propositional logic, the smallest entities dealt with are
    (a) verbs.
    (b) nouns.
    (c) adjectives.
    (d) predicates.
    (e) sentences.

12. Examine Fig. Test 1-2. This is a diagram showing the evolution of a hypothetical system of mathematics. The ellipses with the question marks inside represent
    (a) applications of the rules of logic.
    (b) elementary (undefined) terms.
    (c) lemmas.
    (d) corollaries.
    (e) contradictions.

13. A fallacy is
    (a) a minor theorem that arises directly from a major theorem.
    (b) a minor theorem that is used to help prove a major theorem.
    (c) an axiom disguised as a definition.
    (d) a violation or misapplication of the laws of logic.
    (e) an example of *reductio ad absurdum*.

14. The symbol ∃ is used in
    (a) sentential logic.
    (b) propositional logic.
    (c) predicate logic.
    (d) subjective logic.
    (e) universal logic.

15. Suppose an author is writing a book about the history of a particular region, and in the course of research, the author finds that as the population increased, the number of burglaries per 100,000 people went up in direct proportion. In fact, the two variables tracked along together with uncanny exactness. Which of the following represents sound reasoning, and is therefore a logical conclusion?

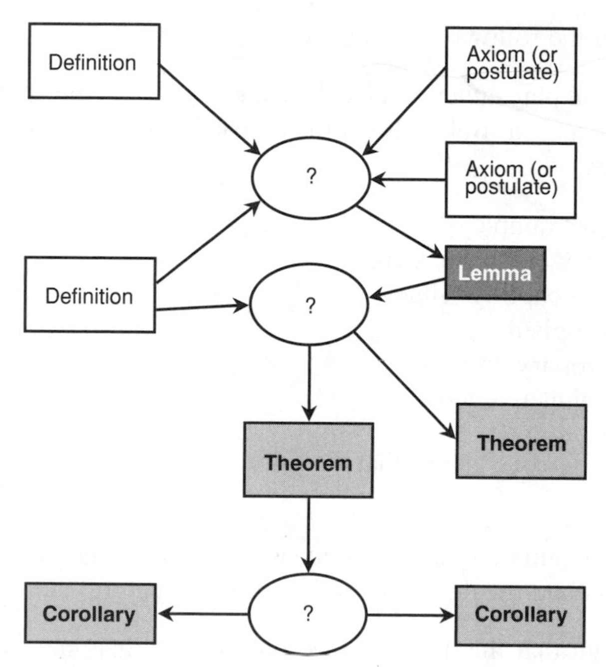

Fig. Test 1-2. Illustration for Part One Test Question 12.

(a) Burglars caused the population of the region to increase.
(b) The population increase caused burglars to move into the region.
(c) Some unknown factor caused the population of the region to increase, and also caused burglars to move into the region.
(d) The population of the region was correlated with the number of burglaries per 100,000 people.
(e) All of the above

16. Which of the following propositions represents an example of the probability fallacy?
    (a) The home team is likely to win tomorrow.
    (b) Either the home team will win tomorrow, or else it will not win.
    (c) Snowstorms cause home teams to win.
    (d) Home-team victories cause snowstorms.
    (e) Home-team victories cause snowstorms, and vice-versa.

17. Consider the following argument:

    - Bob will play either basketball or hockey during the winter high school sports season. Bob will not play basketball. Therefore, Bob will play hockey.

    This is an example of
    (a) denying the antecedent.
    (b) the probability fallacy.
    (c) a syllogism.
    (d) a corollary.
    (e) a conjunctive lemma.

18. Suppose someone claims that the following proposition is a mathematical theorem:

    - All real numbers can be written as fractions, where the numerator is an integer and the denominator is a nonzero natural number.

    You claim that this is not a theorem because it is not true. How many counterexamples must you find in order to show that this proposition is not true?
    (a) None.
    (b) One.
    (c) Two.
    (d) At least a few dozen.
    (e) Infinitely many.

19. In order to prove the falsity of an existential proposition $(\exists x)\ Px$ about a variable x in a universal set U, it is necessary to prove that
    (a) $(\forall x)\ Px$
    (b) $(\forall x) \neg Px$
    (c) Pk, where k is a constant and $k \in U$
    (d) $\neg Pk$, where k is a constant and $k \in U$
    (e) Any of the above

20. Examine Table Test 1-1. Which of the following can be written in the uppermost column header at the extreme right in place of the question mark, making the truth table correct?
    (a) **X & Y & Z**
    (b) $\mathbf{X \vee (Y\ \&\ Z)}$

Table Test 1-1. Table for Part One Test Question 20.

X	Y	Z	?
F	F	F	F
F	F	T	F
F	T	F	F
F	T	T	F
T	F	F	F
T	F	T	F
T	T	F	F
T	T	T	T

(c) **X & (Y ∨ Z)**
(d) **(X ∨ Y) & Z**
(e) **X ∨ Y ∨ Z**

21. Which of the following expressions is not a well-formed formula (wff)? Variables are symbolized as *x* and *y*. Constants are symbolized as *a* and *b*. Predicates are symbolized as R and S.

R*ab*
S*abx*
S*a*R
R*xyab*
S*a*

(a) The first expression
(b) The second expression
(c) The third expression
(d) The fourth expression
(e) The fifth expression

22. Which of the following is not true of mathematical definitions?
(a) They can take the form of SLVC sentences.
(b) They can contain logical equivalences.
(c) They must be proven from axioms, lemmas, and corollaries.

(d) They must be clear and unambiguous.
(e) They are an essential part of any mathematical theory.

23. A truly random sequence of digits must be
    (a) unpredictable; that is, there should be no way to figure out the next digit based on the ones already known.
    (b) rational; that is, it must be expressible as the quotient of an integer and a nonzero natural number.
    (c) such that it is impossible to write it out as a terminating decimal.
    (d) infinite; that is, it must never end.
    (e) All of the above

24. When logic that appears perfectly sound is used to reach an incredible conclusion, it is called
    (a) a paradox.
    (b) a contradiction.
    (c) a theorem.
    (d) inductive reasoning.
    (e) mathematical induction.

25. A corollary is
    (a) a definition that is provable from a set of axioms.
    (b) a definition that contains a contradiction.
    (c) a minor theorem intended to help prove something more important.
    (d) a theorem that arises as a direct consequence of another theorem.
    (e) a subset of the universal set.

26. In the sentence "Jim likes sausage and eggs," the phrase "likes sausage and eggs" is
    (a) the subject.
    (b) the verb.
    (c) the linking verb.
    (d) the object.
    (e) the predicate.

27. Consider the following compound sentence, where X, Y, and Z are variables representing whole sentences:

$$X \& (Y \vee Z) \Leftrightarrow (X \& Y) \vee (X \& Z)$$

This is a statement of

(a) the associative law of disjunction with respect to conjunction.
(b) the associative law of conjunction with respect to disjunction.
(c) the distributive law of conjunction with respect to disjunction.
(d) the commutative law of conjunction with respect to disjunction.
(e) the commutative law of disjunction with respect to conjunction.

28. A wff can contain

(a) both conjunction and disjunction.
(b) both negation and conjunction.
(c) both constants and variables.
(d) more than one predicate.
(e) Any of the above

29. If someone comes to a broad-based conclusion based on insufficient information, that person commits

(a) *reductio ad absurdum.*
(b) mathematical induction.
(c) a paradox.
(d) implication reversal.
(e) a hasty generalization.

30. Suppose a strange new disease appears. Five doctors all disagree about its cause. Doctor V says, "I think it is caused by a virus." Doctor W says, "No, I have good evidence to suggest that it is caused by bacteria." Doctor X says, "Most likely, it is caused by poor nutrition." Doctor Y says, "I suspect it is caused by air pollution." Doctor Z says, "I think you are all wrong. I have a feeling it is caused by an as-yet unknown agent." Which of these doctors commits a logical fallacy in making his or her statement?

(a) Doctor V
(b) Doctor W
(c) Doctor X
(d) Doctor Y
(e) Doctor Z

31. A form of logic that involves a continuous span of truth values, for example a range from 0 (totally false) through 1 (neutral) to 2 (totally true), is known as

(a) dichotomous logic.
(b) fuzzy logic.
(c) inductive logic.
(d) numerical logic.
(e) precision logic.

32. In the sentence “I am an amateur radio operator,” the word “am” is
    (a) the subject.
    (b) the complement.
    (c) the linking verb.
    (d) the object.
    (e) the predicate.

33. In a logical equivalence, the double-shafted, double-headed arrow can be replaced by the word or words
    (a) “and.”
    (b) “if.”
    (c) “if and only if.”
    (d) “implies.”
    (e) “or.”

34. The associative law of disjunction is concerned with
    (a) the order in which sentences are stated when connected by disjunctions.
    (b) the way in which sentences are grouped when connected by disjunctions.
    (c) the negation of a compound sentence containing disjunctions.
    (d) the negation of the individual sentences in a compound sentence containing disjunctions.
    (e) All of the above

35. The symbol $\forall$ can be replaced with the word or phrase
    (a) “for none.”
    (b) “for one.”
    (c) “for some.”
    (d) “for every.”
    (e) “for infinitely many.”

36. A sentence that is a logical disjunction of two or more component sentences is true

(a) if and only if none of its components is true.
(b) if and only if at least one of its components is true.
(c) if and only if all of its components are true.
(d) if and only if at least one of its components is false.
(e) no matter what.

37. Which of the following (a), (b), (c), or (d), if any, is never done in the development a rigorous mathematical theory?
(a) Axioms and definitions are used to prove a lemma.
(b) Axioms and a lemma are used to prove a definition.
(c) Axioms, definitions, and a lemma are used to prove a theorem.
(d) Axioms, definitions, and a theorem are used to prove a corollary.
(e) All of the above (a), (b), (c), and (d) are commonly done in the development of a rigorous mathematical theory.

38. Once a proposition has been proved in a mathematical system, that proposition becomes
(a) an axiom.
(b) a definition.
(c) a logical equivalence.
(d) a contradiction.
(e) a theorem.

39. Consider the following compound sentence, where X, Y, and Z are variables representing whole sentences:

$$(X \,\&\, Y) \,\&\, Z \Leftrightarrow X \,\&\, (Y \,\&\, Z)$$

This is a statement of
(a) the associative law of conjunction.
(b) the associative law of disjunction.
(c) the distributive law of conjunction.
(d) the commutative law of disjunction.
(e) the commutative law of conjunction.

40. A sentence of the form "If P, then Q" is false
(a) if and only if P is true and Q is true.
(b) if and only if P is true and Q is false.
(c) if and only if P is false and Q is true.
(d) if and only if P is false and Q is false.
(e) no matter what.

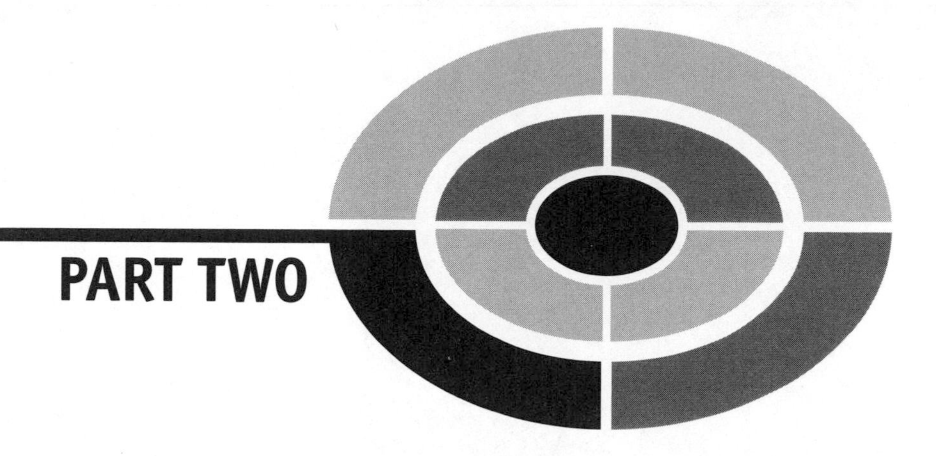

PART TWO

Proofs in Action

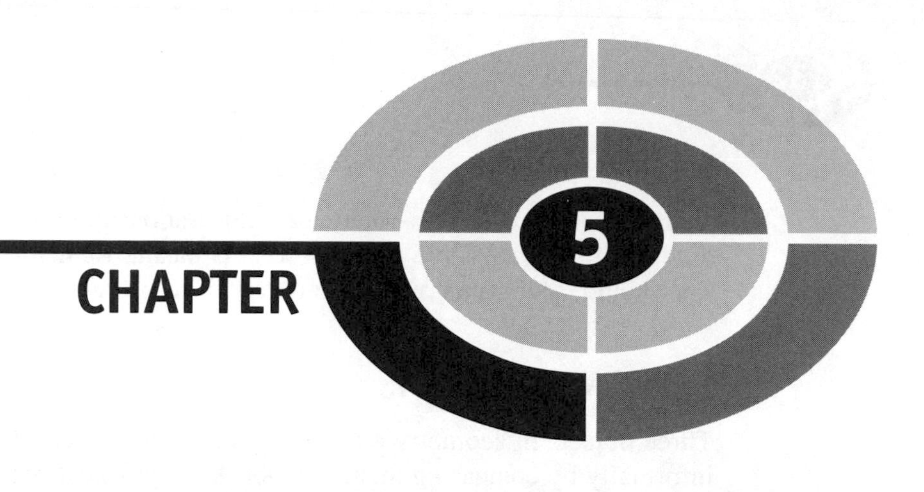

Some Theoretical Geometry

The material in this chapter involves basic *plane geometry*, also called *Euclidean geometry*. We'll work with points, lines, line segments, rays, angles, and triangles. This is not a geometry course, but it will show you how definitions and axioms are formulated, and how they can be used to prove theorems.

Don't hurry through this material. Let it "soak in" slowly. As you go through this text and measure your progress in hours per page, perhaps you will be reminded of the martial arts, where you are required go through certain routines over and over until they become second nature. The proving of mathematical theorems is like that; it is an art that requires practice and repetition.

Some Definitions

Let's get started by stating some definitions. We must know what we're talking about before we can prove any propositions! Some of these definitions are accompanied by illustrations. In other cases, illustrations are not provided, but you can draw them if it helps you envision what is meant. Remember that a good definition should stand on its own, and not rely on any drawings to convey its meaning.

ELEMENTARY TERMS

Three objects in geometry are undefined, or elementary. We can describe them informally by comparing them with idealized physical objects.

- A *point* is like a ball with a radius of 0, or a brick that measures 0 units along each edge. A point has a location or position that can be specified with absolute precision, but it has no volume or mass. All points are *0-dimensional* (0D).
- A *line* is like an infinitely fine, infinitely long, perfectly straight wire. It extends forever in two opposite directions. A line has a position and an orientation that can be specified with absolute precision, but it has no volume or mass. All lines are *1-dimensional* (1D).
- A *plane* is like an infinitely thin, perfectly flat pane of glass that goes on forever without any edges. A plane has a position and orientation that can be specified with absolute precision, but it has no volume or mass. All planes are *2-dimensional* (2D).

A line has no end points. A plane has no edges. These properties, along with the fact that a point has position but no dimension, make the point, the line, and the plane strange indeed! You will never whack a point with a golf club, slice a chunk of cheese with a line, or sail an ice boat across a plane. Points, lines, and planes are not material things, and yet they are exactly what we think they are!

LINE SEGMENT

Let P and Q be distinct points that lie on a line L. The *closed line segment PQ* is the set of all points on L between, and including, points P and Q. This is illustrated in Fig. 5-1A. When you hear or read the term *line segment*, it is meant to refer to a closed line segment unless otherwise specified.

HALF-OPEN LINE SEGMENT

Let P and Q be distinct points that lie on a line L. The *half-open line segment PQ* can be the set of all points on L between P and Q but not including P, or the set of all points on L between P and Q but not including Q. These are illustrated in

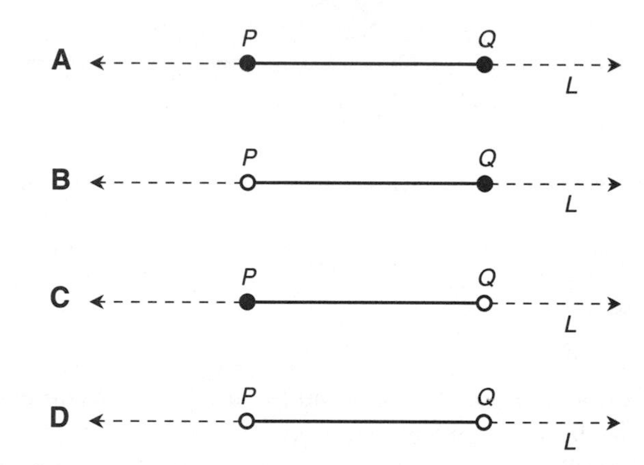

Fig. 5-1. At A, the closed line segment *PQ*. At B and C, half-open line segments *PQ*. At D, the open line segment *PQ*.

Figs. 5-1B and 5-1C. The *excluded end points* are shown as open circles, and the *included end points* are shown as solid dots.

OPEN LINE SEGMENT

Let *P* and *Q* be distinct points that lie on a line *L*. The *open line segment PQ* is the set of all points on *L* between, but not including, points *P* and *Q*, as shown in Fig. 5-1D.

LENGTH OF LINE SEGMENT

Let *PQ* be a closed, half-open, or open line segment defined by distinct end points *P* and *Q*. The *length* of line segment *PQ* is the shortest possible distance between points *P* and *Q*. The length of line segment *QP* is considered equal to (not the negative of) the length of line segment *PQ*. That is, it does not matter in which direction the length of a line segment is expressed.

CLOSED-ENDED RAY

Let *P* and *Q* be distinct points that lie on a line *L*. The *closed-ended ray PQ*, also called the *closed-ended half-line PQ*, consists of the set of all points on *L* that lie on the side of point *P* that contains point *Q*, including point *P* itself. This is illus-

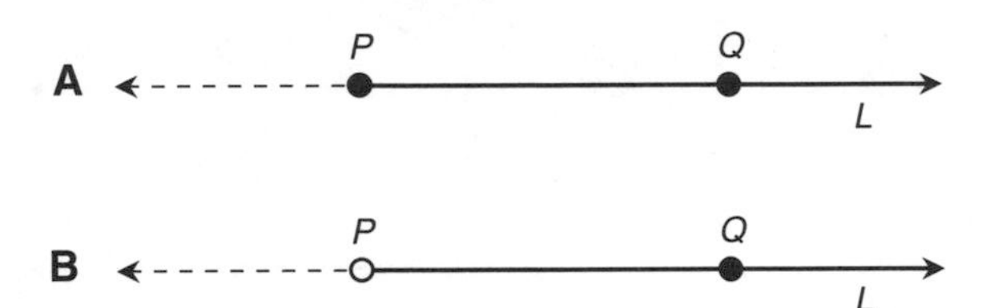

Fig. 5-2. At A, the closed-ended ray *PQ*. At B, the open-ended ray *PQ*.

trated in Fig. 5-2A. When you hear or read the term *ray*, it is meant to refer to a closed-ended ray unless otherwise specified.

OPEN-ENDED RAY

Let *P* and *Q* be distinct points that lie on a line *L*. The *open-ended ray PQ*, also called the *open-ended half-line PQ*, consists of the set of all points on *L* that lie on the side of point *P* that contains point *Q*, but not including point *P* itself. This is illustrated in Fig. 5-2B.

POINT OF INTERSECTION

Let *PQ* be a line, line segment, or ray defined by distinct points *P* and *Q*. Let *RS* be a line, line segment, or ray defined by points *R* and *S*, different from points *P* and *Q*. Suppose *T* is a point that lies on both *PQ* and *RS*. Then *T* is a *point of intersection* between *PQ* and *RS*. Also, *PQ* and *RS* are said to *intersect* at point *T*.

COLLINEAR POINTS

Let $P_1, P_2, P_3, \ldots, P_n$ be *mutually distinct* points (that means no two of the points coincide). Then $P_1, P_2, P_3, \ldots, P_n$ are *collinear* if and only if they all lie on a single line.

COPLANAR POINTS

Let $P_1, P_2, P_3, \ldots, P_n$ be mutually distinct points. Then $P_1, P_2, P_3, \ldots, P_n$ are *coplanar* if and only if they all lie in a single plane.

COINCIDENT LINES

Let *P*, *Q*, *R*, and *S* be distinct points. Line *PQ*, defined by points *P* and *Q*, and line *RS*, defined by points *R* and *S*, are *coincident lines* if and only if points *P*, *Q*, *R*, and *S* are collinear.

COLLINEAR LINE SEGMENTS AND RAYS

Let *P*, *Q*, *R*, and *S* be distinct points. Let *PQ* represent a closed, half-open, or open line segment or ray defined by points *P* and *Q*. Let *RS* represent a closed, half-open, or open line segment or ray defined by points *R* and *S*. Then *PQ* and *RS* are *collinear* if and only if points *P*, *Q*, *R*, and *S* are collinear.

TRANSVERSAL

Let *P*, *Q*, *R*, and *S* be distinct points. Let *PQ* be the line defined by points *P* and *Q*. Let *RS* be the line defined by points *R* and *S*. Suppose that line *PQ* and line *RS* lie in the same plane, but are not coincident. Let *L* be a line that intersects both line *PQ* and line *RS*. Line *L* is said to be a *transversal* of lines *PQ* and *RS*.

PARALLEL LINES

Let *P*, *Q*, *R*, and *S* be distinct points. Let *PQ* be the line defined by points *P* and *Q*. Let *RS* be the line defined by points *R* and *S*. Suppose that lines *PQ* and *RS* lie in the same plane, but are not coincident. Then *PQ* and *RS* are *parallel lines* if and only if there exists no point *T* such that *T* is a point of intersection between line *PQ* and line *RS*.

PARALLEL LINE SEGMENTS

Let *P*, *Q*, *R*, and *S* be distinct points. Let line *PQ* be the line defined by points *P* and *Q*. Let line segment *PQ* be a closed, half-open, or open line segment contained in line *PQ*, with end points *P* and *Q*. Let *RS* be the line defined by points *R* and *S*. Suppose that lines *PQ* and *RS* lie in the same plane, but are not coincident. Let line segment *RS* be a closed, half-open, or open line segment contained in line *RS*, with end points *R* and *S*. Then line segments *PQ* and *RS* are *parallel line segments* if and only if lines *PQ* and *RS* are parallel lines.

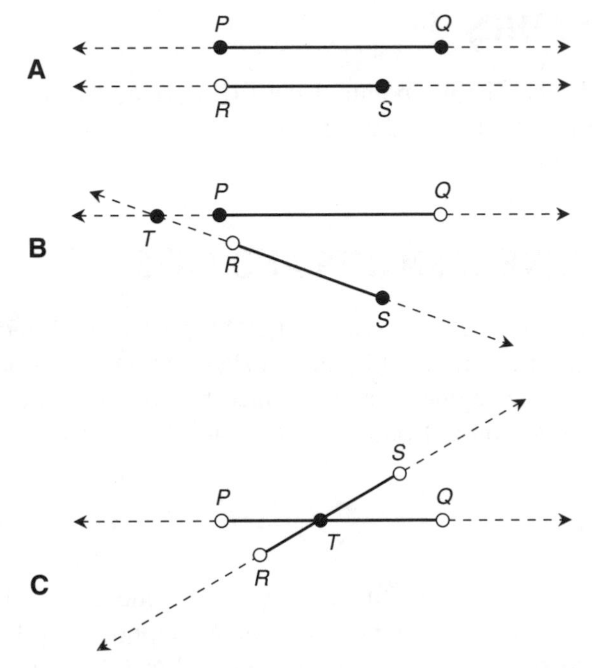

Fig. 5-3. At A, line segments *PQ* and *RS* are parallel, because lines *PQ* and *RS* lie in the same plane and do not intersect. At B and C, line segments *PQ* and *RS* are not parallel, because lines *PQ* and *RS* intersect at a point *T*.

In Fig. 5-3, drawing A shows an example of parallel line segments, one closed and the other half-open. Drawing B shows an example of two half-open line segments that are not parallel. Drawing C shows an example of two open line segments that are not parallel. In the situation of Fig. 5-3A, there exists no point of intersection *T* common to both lines. In the situation of Fig. 5-3B, there is a point *T* common to both lines, although *T* does not lie on either line segment. In the situation of Fig. 5-3C, there exists a point *T* common to both lines, and *T* lies on both line segments.

PARALLEL RAYS

Let *P*, *Q*, *R*, and *S* be distinct points. Let *PQ* be the line defined by points *P* and *Q*. Let ray *PQ* be a closed-ended or open-ended ray contained in line *PQ*, with end point *P*. Let *RS* be the line defined by points *R* and *S*. Suppose that lines *PQ* and *RS* lie in the same plane, but are not coincident. Let ray *RS* be a closed-ended

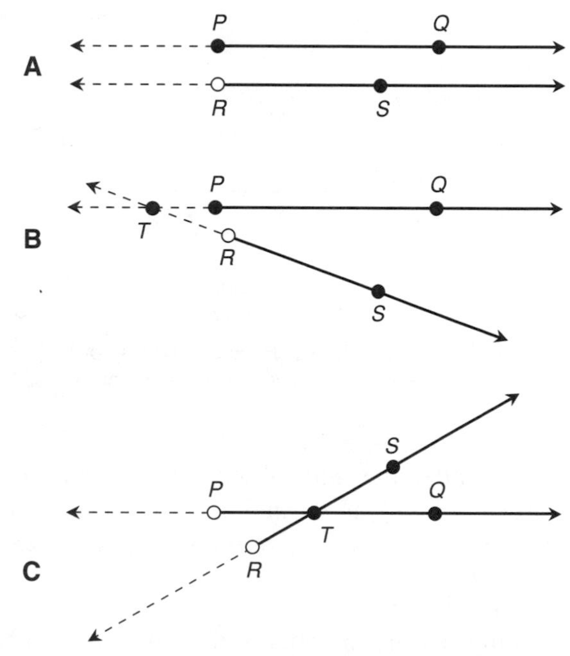

Fig. 5-4. At A, rays *PQ* and *RS* are parallel, because lines *PQ* and *RS* lie in the same plane and do not intersect. At B and C, rays *PQ* and *RS* are not parallel, because lines *PQ* and *RS* intersect at a point *T*.

or open-ended ray contained in line *RS*, with end point *R*. Then rays *PQ* and *RS* are *parallel rays* if and only if lines *PQ* and *RS* are parallel lines.

In Fig. 5-4, drawing A shows an example of parallel rays, one closed-ended and the other open-ended. Drawing B shows an example of two rays, one closed-ended and the other open-ended, that are not parallel. Drawing C shows an example of two rays, both open-ended, that are not parallel. In the situation of Fig. 5-4A, there exists no point of intersection *T* common to both lines. In the situation of Fig. 5-4B, there is a point *T* common to both lines, although *T* does not lie on either ray. In the situation of Fig. 5-4C, there exists a point *T* common to both lines, and *T* lies on both rays.

ANGLE

Let *P*, *Q*, and *R* be distinct points. Let *QP* and *QR* be rays or line segments, both of which have the same end point, *Q*. Then the two rays or line segments and their common end point constitute an *angle* denoted $\angle PQR$. Rays or line seg-

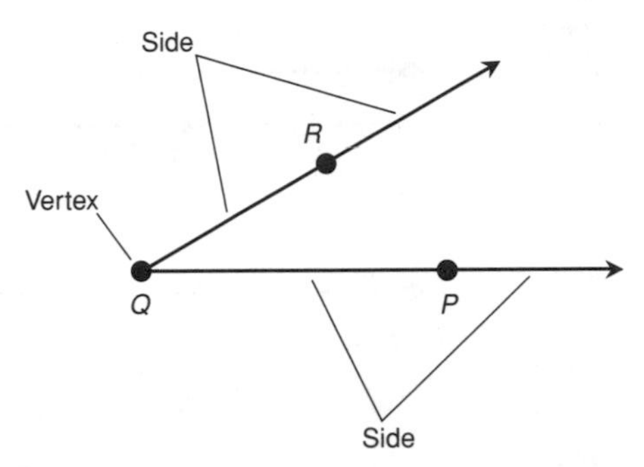

Fig. 5-5. An angle is defined by two rays or line segments that share an end point.

ments QP and QR are called the *sides* of $\angle PQR$, and point Q is called the *vertex* of $\angle PQR$. An example is shown in Fig. 5-5. Unless otherwise specified, angles are expressed by counterclockwise rotation around the vertex.

PROBLEM 5-1

The preceding definition actually describes two different angles with the same vertex. What are they?

SOLUTION 5-1

The two rays described, and shown in Fig. 5-5, can define two different angles! One of the angles, which is the one you're most likely to imagine when you look at the drawing, goes counterclockwise "the short way around" from ray QP to ray QR. The other angle, which is also represented in the definition but which you're less likely to imagine, goes counterclockwise "the long way around" from ray QR to ray QP. It can be denoted $\angle RQP$.

PROBLEM 5-2

What about angles that are defined as going clockwise, rather than counterclockwise? Are they legitimate angles, too? How can these be described in the situation shown in Fig. 5-5?

SOLUTION 5-2

Angles can be expressed as going clockwise instead of counterclockwise, but this is the unconventional method of defining angle direction. An angle going clockwise can be considered the negative of the angle having the same sides going counterclockwise. In the scenario of Fig. 5-5 going clockwise, $\angle RQP$ goes the short way around, and $\angle PQR$ goes the long way around.

MEASURE OF ANGLE

Let P, Q, and R be distinct points. Let QP and QR be distinct, closed-ended rays or closed line segments, both of which have the same end point, Q. Then the *measure* of $\angle PQR$, denoted $m\angle PQR$, is an expression of the portion of a complete revolution described by $\angle PQR$. The *measure in degrees* of $\angle PQR$, symbolized $m^\circ\angle PQR$, is the portion of a complete revolution described by $\angle PQR$, multiplied by 360. The measure of an angle is often denoted as a variable in lowercase English or Greek, such as x, y, θ, or ϕ.

STRAIGHT ANGLE

Let P, Q, and R be distinct points. Let QP and QR be two rays or line segments, both of which have the same end point, Q, and which define $\angle PQR$. Then $\angle PQR$ is a *straight angle* if and only if points P, Q, and R are collinear and point Q is between point P and point R.

STRAIGHT ANGLE (ALTERNATE DEFINITION)

An angle is a *straight angle* if and only if its measure is ½ of a complete revolution.

STRAIGHT ANGLE (SECOND ALTERNATE DEFINITION)

An angle is a *straight angle* if and only its measure is equal to 180°.

SUPPLEMENTARY ANGLES

Two angles are *supplementary angles* if and only if the sum of their measures is equal to the measure of a straight angle.

RIGHT ANGLE

Let P, Q, R, S, and T be distinct points, all of which lie in the same plane, but not all of which are collinear. Consider ray TP, ray TQ, ray TR, and ray TS. Suppose the following are true:

- The rays define $\angle PTQ$, $\angle QTR$, $\angle RTS$, and $\angle STP$
- All four angles $\angle PTQ$, $\angle QTR$, $\angle RTS$, and $\angle STP$ have equal measure

Then each of the angles $\angle PTQ$, $\angle QTR$, $\angle RTS$, and $\angle STP$ is a *right angle*.

RIGHT ANGLE (ALTERNATE DEFINITION)

An angle is a *right angle* if and only if its measure is ¼ of a complete revolution.

RIGHT ANGLE (SECOND ALTERNATE DEFINITION)

An angle is a *right angle* if and only its measure is equal to 90º.

COMPLEMENTARY ANGLES

Two angles are *complementary angles* if and only if the sum of their measures is equal to the measure of a right angle.

PERPENDICULAR LINES, LINE SEGMENTS, AND RAYS

Let P, Q, R, and S be distinct points. A line, line segment, or ray PQ is *perpendicular* to a line, line segment, or ray RS if and only if both of the following are true:

- Line, line segment, or ray PQ intersects line, line segment, or ray RS at one and only one point T
- One of the angles at point T, formed by the intersection of line PQ and line RS, is a right angle

TRIANGLE

Let P, Q, and R be three distinct points. Consider the closed line segment PQ, the closed line segment QR, and the closed line segment RP. These three line segments, along with the points P, Q, and R, constitute a *triangle* denoted ΔPQR. Line segment PQ, line segment QR, and line segment RP are called the *sides* of ΔPQR. Points P, Q, and R are called the *vertices* of ΔPQR. The three angles,

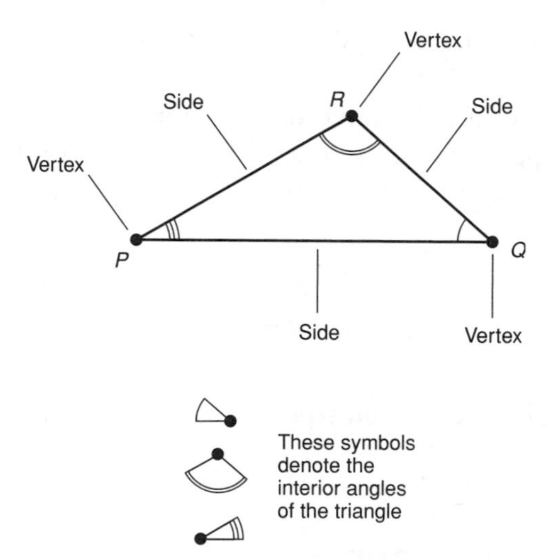

Fig. 5-6. A triangle is defined by three points and the line segments that connect them.

$\angle RQP$, $\angle PRQ$, and $\angle QPR$, are called the *interior angles* of ΔPQR. Fig. 5-6 illustrates an example.

The interior angles of a triangle can be identified by drawing arcs centered on their vertices and connecting their sides, as shown. Arrows can be used to specify the rotational sense of the angle. If no arrows are shown, then the rotational sense should be considered as counterclockwise. If specific vertex points are not important, triangles can be named with italicized, uppercase letters from the latter part of the alphabet, such as *T*, *U*, and *V*.

INCLUDED ANGLE

In a triangle, an *included angle* is the interior angle whose vertex is the point where two specified sides of a triangle intersect. For example, in Fig. 5-6, $\angle QPR$ is the included angle between the sides represented by line segments *PQ* and *PR*.

INCLUDED SIDE

In a triangle, an *included side* is a side whose end points constitute the vertices of two specified angles. For example, in Fig. 5-6, line segment *PQ* is the included side between $\angle RQP$ and $\angle QPR$.

ISOSCELES TRIANGLE

Imagine a triangle defined by three distinct points P, Q, and R. Let p be the length of the side opposite point P. Let q be the length of the side opposite point Q. Let r be the length of the side opposite point r. Suppose any of the following equations hold:

$$p = q$$
$$q = r$$
$$p = r$$

This kind of triangle is called an *isosceles triangle*. It has two sides of equal length.

EQUILATERAL TRIANGLE

Imagine a triangle defined by three distinct points P, Q, and R. Let p be the length of the side opposite point P. Let q be the length of the side opposite point Q. Let r be the length of the side opposite point R. Suppose the following equation holds:

$$p = q = r$$

This kind of triangle is called an *equilateral triangle*. All three sides are of equal length.

RIGHT TRIANGLE

A triangle is called a *right triangle* if and only if one of its interior angles is a right angle.

Similar and Congruent Triangles

Here are some definitions that apply especially to triangles. In our theorem-proving exercises (we're getting there!), we will be discussing *similarity* and *congruence*, and in particular, a property called *direct congruence*. There are four important definitions to consider here. They are a little tricky, so pay attention!

DIRECT SIMILARITY

Here is a formal definition, which you can visualize by looking at Fig. 5-7A:

- Let P, Q, and R be distinct points. Let ΔPQR be a triangle defined by proceeding counterclockwise from point P to point Q, from point Q to point R, and from point R to point P. Let p be the length of the side opposite point P. Let q be the length of the side opposite point Q. Let r be the length of the side opposite point R. Let S, T, and U be distinct points. Suppose that points P, Q, R, S, T, and U are all coplanar. Let ΔSTU be a triangle distinct from ΔPQR, proceeding counterclockwise from point S to point T, from point T to point U, and from point U to point S. Let s be the length of the side opposite point S. Let t be the length of the side opposite point T. Let u be the length of the side opposite point U. Then ΔPQR and ΔSTU are *directly similar* if and only if the lengths of their corresponding sides, as we proceed in the same direction around either triangle, are in a constant ratio; that is, if and only if $p/s = q/t = r/u$. In addition, ΔPQR and ΔSTU are directly similar if and only if the counterclockwise measures of their corresponding angles, as we proceed in the same direction around either triangle, are equal; that is, if and only if $m\angle QPR = m\angle TSU$, $m\angle PRQ = m\angle SUT$, and $m\angle RQP = m\angle UTS$.

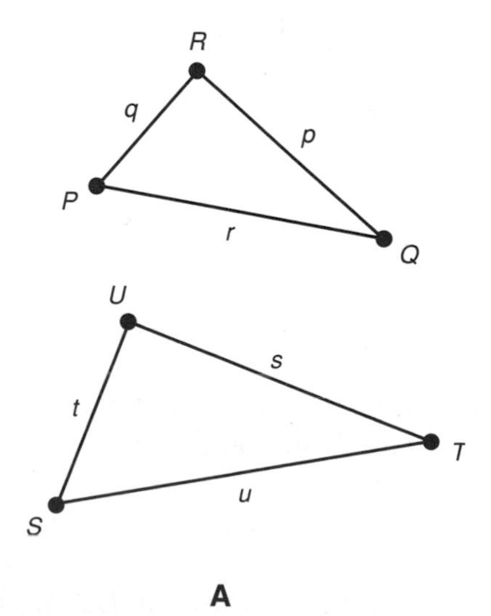

Fig. 5-7A. Directly similar triangles.

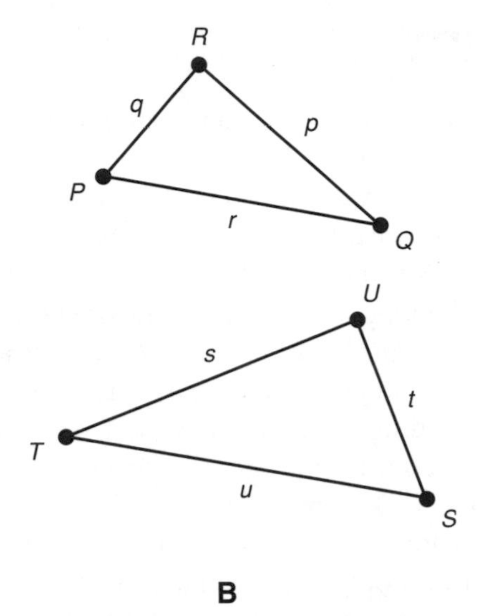

Fig. 5-7B. Inversely similar triangles.

An alternative, but informal, way to describe direct similarity is to imagine that, if one triangle is enlarged or reduced by just the right amount and then rotated clockwise or counterclockwise to just the right extent (but not flipped over), one triangle can be pasted on top of the other so they coincide.

The direct similarity symbol looks like a wavy minus sign (~). If distinct triangles ΔPQR and ΔSTU are directly similar, we symbolize the fact like this:

$$\Delta PQR \sim \Delta STU$$

INVERSE SIMILARITY

Here is a formal definition, which you can visualize by looking at Fig. 5-7B:

- Let P, Q, and R be distinct points. Let ΔPQR be a triangle defined by proceeding counterclockwise from point P to point Q, from point Q to point R, and from point R to point P. Let p be the length of the side opposite point P. Let q be the length of the side opposite point Q. Let r be the length of the side opposite point R. Let S, T, and U be distinct points. Suppose that points P, Q, R, S, T, and U are all coplanar. Let ΔSTU be a triangle distinct from ΔPQR, proceeding clockwise from point S to point T, from point T to point U, and from point U to point S. Let s be the length of the side opposite point S. Let t be the length of the side opposite point T. Let u be the length of the side opposite point U. Then ΔPQR and ΔSTU are *inversely similar* if and only if the lengths of their corresponding sides, as we proceed in opposite directions around the triangles, are in a constant ratio; that is, if and only if $p/s = q/t = r/u$. In addition, ΔPQR and ΔSTU are inversely similar if and only if the counterclockwise measures of their corresponding angles, as we proceed in opposite directions around the triangles, are equal; that is, if and only if $m\angle QPR = m\angle UST$, $m\angle PRQ = m\angle TUS$, and $m\angle RQP = m\angle STU$.

An alternative, but informal, way to describe inverse similarity is to imagine that, if one triangle is flipped over, enlarged or reduced by just the right amount, and finally rotated clockwise or counterclockwise to just the right extent, one triangle can be pasted on top of the other so they coincide.

The inverse similarity symbol is not universally agreed-upon. Let's use a wavy minus sign followed by a regular minus sign (~–). If distinct triangles ΔPQR and ΔSTU are inversely similar, then, we can symbolize the fact like this:

$$\Delta PQR \sim\!- \Delta STU$$

When you read or hear that two triangles are *similar triangles*, it is not necessarily clear what the author means! Sometimes the term *similarity* is applied only to direct similarity, but in some texts it can refer to either direct similarity or inverse similarity. We can avoid that sort of confusion by using the full terminology all the time.

DIRECT CONGRUENCE

Here is a formal definition, which you can visualize by looking at Fig. 5-8A:

- Let P, Q, and R be distinct points. Let ΔPQR be a triangle defined by proceeding counterclockwise from point P to point Q, from point Q to point R, and from point R to point P. Let p be the length of the side opposite point P. Let q be the length of the side opposite point Q. Let r be the length of the side opposite point R. Let S, T, and U be distinct points. Suppose that points P, Q, R, S, T, and U are all coplanar. Let ΔSTU be a triangle distinct from ΔPQR, proceeding counterclockwise from point S to point T, from

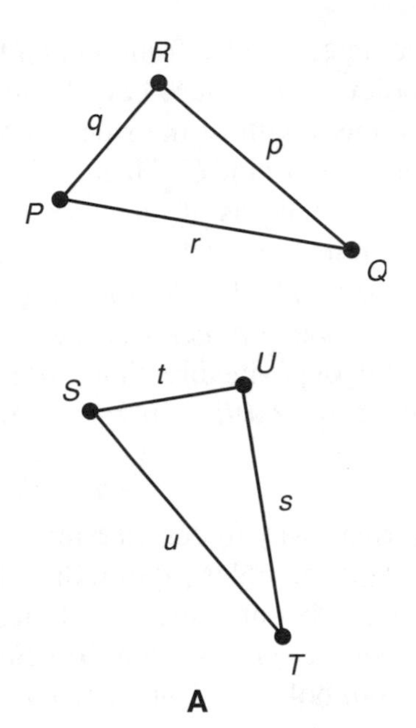

Fig. 5-8A. Directly congruent triangles.

point T to point U, and from point U to point S. Let s be the length of the side opposite point S. Let t be the length of the side opposite point T. Let u be the length of the side opposite point U. Then ΔPQR and ΔSTU are *directly congruent* if and only if the lengths of their corresponding sides, as we proceed in the same direction around either triangle, are equal; that is, if and only if $p = s$, $q = t$, and $r = u$. In addition, if ΔPQR and ΔSTU are directly congruent, then the counterclockwise measures of their corresponding angles, as we proceed in the same direction around either triangle, are equal; that is, $m\angle QPR = m\angle TSU$, $m\angle PRQ = m\angle SUT$, and $m\angle RQP = m\angle UTS$.

An alternative, but informal, way to describe direct congruence is to imagine that, if one triangle is rotated clockwise or counterclockwise to just the right extent (but not flipped over), one triangle can be pasted on top of the other, and they will coincide.

The direct congruence symbol looks like a triple-barred equals sign (≡). If distinct triangles ΔPQR and ΔSTU are directly congruent, we symbolize the fact like this:

$$\Delta PQR \equiv \Delta STU$$

INVERSE CONGRUENCE

Here is a formal definition, which you can visualize by looking at Fig. 5-8B:

- Let P, Q, and R be distinct points. Let ΔPQR be a triangle defined by proceeding counterclockwise from point P to point Q, from point Q to point R, and from point R to point P. Let p be the length of the side opposite point P. Let q be the length of the side opposite point Q. Let r be the length of the side opposite point R. Let S, T, and U be distinct points. Suppose that points P, Q, R, S, T, and U are all coplanar. Let ΔSTU be a triangle distinct from ΔPQR, proceeding clockwise from point S to point T, from point T to point U, and from point U to point S. Let s be the length of the side opposite point S. Let t be the length of the side opposite point T. Let u be the length of the side opposite point U. Then ΔPQR and ΔSTU are *inversely congruent* if and only if the lengths of their corresponding sides, as we proceed in opposite directions around the triangles, are equal; that is, if and only if $p = s$, $q = t$, and $r = u$. In addition, if ΔPQR and ΔSTU are inversely congruent, then the counterclockwise measures of their corresponding angles, as we proceed in opposite directions around the triangles, are equal; that is, $m\angle QPR = m\angle UST$, $m\angle PRQ = m\angle TUS$, and $m\angle RQP = m\angle STU$.

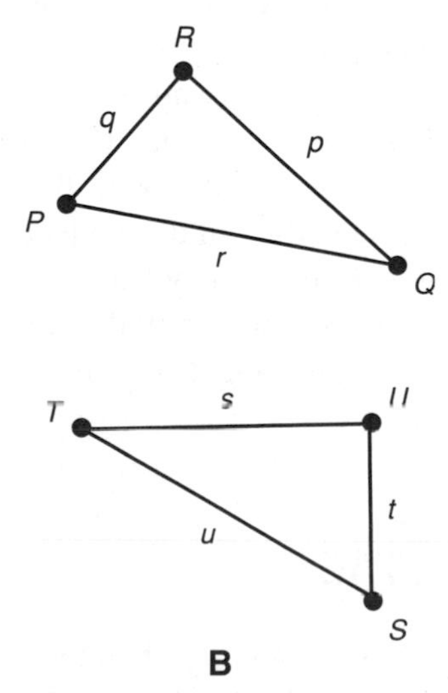

Fig. 5-8B. Inversely congruent triangles.

An alternative, but informal, way to describe inverse congruence is to imagine that, if one triangle is flipped over and then rotated clockwise or counterclockwise to just the right extent, one triangle can be pasted on top of the other, and they will coincide.

The inverse congruence symbol is not universally agreed-upon. Let's use a triple-barred equals sign followed by a minus sign (≡–). If distinct triangles ΔPQR and ΔSTU are inversely congruent, then, we can symbolize the fact like this:

$$\Delta PQR \equiv\!- \Delta STU$$

When you read or hear that two triangles are *congruent triangles*, it is not necessarily clear what the author means! Sometimes the term *congruence* is applied only to direct congruence, but in some texts it can refer to either direct congruence or inverse congruence. As with similarity, let's use the full terminology all the time!

TWO CRUCIAL FACTS

Here are two important things you should remember about directly congruent triangles. This will help reinforce, in your mind, the meaning of this term.

- If two triangles are directly congruent, then their corresponding sides have equal lengths as you proceed around both triangles in the same direction. The converse of this is also true. If two triangles have corresponding sides with equal lengths as you proceed around them both in the same direction, then the two triangles are directly congruent.
- If two triangles are directly congruent, then their corresponding interior angles (that is, the interior angles opposite the corresponding sides) have equal measures as you proceed around both triangles in the same direction. The converse of this is not necessarily true. It is possible for two triangles to have corresponding interior angles with equal measures when you proceed around them both in the same direction, and yet the two triangles are not directly congruent.

TWO MORE CRUCIAL FACTS

Here are two "mirror images" of the facts just stated. They concern triangles that are inversely congruent. Note the subtle differences in the wording!

- If two triangles are inversely congruent, then their corresponding sides have equal lengths as you proceed around the triangles in opposite directions. The converse of this is also true. If two triangles have corresponding sides with equal lengths as you proceed around them in opposite directions, then the two triangles are inversely congruent.
- If two triangles are inversely congruent, then their corresponding interior angles have equal measures as you proceed around the triangles in opposite directions. The converse of this is not necessarily true. It is possible for two triangles to have corresponding interior angles with equal measures as you proceed around them in opposite directions, and yet the two triangles are not inversely congruent.

PROBLEM 5-3

State a special property of equilateral triangles, and state why it is true.

SOLUTION 5-3

Any two equilateral triangles are directly similar. This is because the ratio of their corresponding sides, proceeding counterclockwise around both triangles, is a constant. Suppose ΔPQR is an equilateral triangle with sides of length m. Suppose ΔSTU is an equilateral triangle with sides of length n. Then the ratio of the lengths of their corresponding sides, proceeding counterclockwise around ΔPQR from point P and counterclockwise around ΔSTU from point S, is always m/n. The triangles are therefore directly similar by definition.

PROBLEM 5-4

State another special property of equilateral triangles, and state why it is true.

SOLUTION 5-4

Any two equilateral triangles are inversely similar. This is because the ratio of their corresponding sides, proceeding counterclockwise around one of them and clockwise around the other, is a constant. Suppose ΔPQR is an equilateral triangle with sides of length m. Suppose ΔSTU is an equilateral triangle with sides of length n. Then the ratio of the lengths of their corresponding sides, proceeding counterclockwise around ΔPQR from point P and clockwise around ΔSTU from point S, is always m/n. (In fact, the ratio of the length of *any* side of ΔPQR to the length of *any* side of ΔSTU is equal to m/n!) The triangles are therefore inversely similar by definition.

Some Axioms

We are now armed with a good supply of definitions. We still need axioms if we want to prove anything. Here are several. The first four are Euclid's first, second, fourth, and fifth postulates. They were stated and illustrated in Chapter 3, but they are worth repeating here, with a few changes to make them more relevant to this chapter.

THE TWO-POINT AXIOM

Any two distinct points P and Q can be connected by a straight line segment.

THE EXTENSION AXIOM

Any straight line segment, defined by distinct points P and Q, can be extended indefinitely and continuously to form a straight line defined by points P and Q.

THE RIGHT ANGLE AXIOM

All right angles have the same measure.

THE PARALLEL AXIOM

Let L be a straight line, and let P be some point not on L. Then there exists one and only one straight line M that passes through point P, such that line M is parallel to line L.

THE SIDE-SIDE-SIDE (SSS) AXIOM

Let T and U be triangles. Suppose T and U have corresponding sides of identical lengths as your proceed around the triangles in the same direction. Then T and U are directly congruent. Also, if T and U are directly congruent triangles, then T and U have corresponding sides of identical lengths as you proceed around the triangles in the same direction. (Fig. 5-9 can help you visualize this. In this example, $a = d$, $b = e$, and $c = f$.)

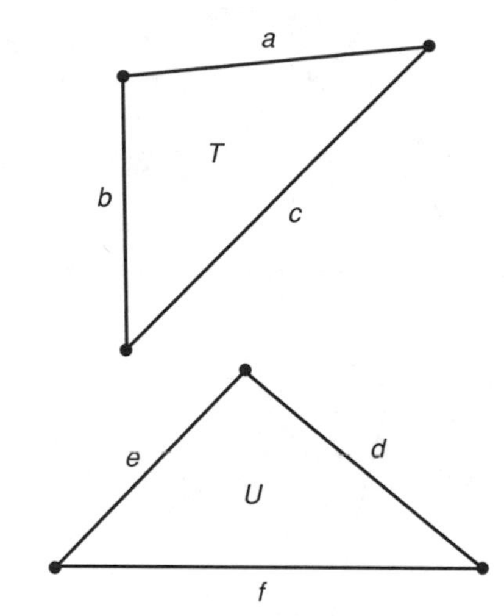

Fig. 5-9. The side-side-side (SSS) axiom. Triangles T and U are directly congruent if and only if $a = d$, $b = e$, and $c = f$.

As an extension of this axiom, suppose T and U have corresponding sides of identical lengths as you proceed around the triangles in opposite directions. Then T and U are inversely congruent. Also, if T and U are inversely congruent triangles, then T and U have corresponding sides of identical lengths as your proceed around the triangles in opposite directions.

Do you think this axiom replicates the definitions of direct and inverse congruence given above? Well, if so, you are right! We state this axiom here because, in some texts, the definitions of direct and inverse congruence are less precise, stating only the general notions about size and shape.

THE SIDE-ANGLE-SIDE (SAS) AXIOM

Let T and U be triangles. Suppose T and U have two pairs of corresponding sides of equal lengths as you proceed around the triangles in the same direction. Also, suppose the included angles between those corresponding sides have identical measures. Then T and U are directly congruent. Also, if T and U are directly congruent triangles, then T and U have two pairs of corresponding sides of equal lengths as you proceed around the triangles in the same direction, and the included angles between those corresponding sides have identical measures. (Fig. 5-10 can help you visualize this. In this example, $a = c$, $b = d$, and $x = y$.)

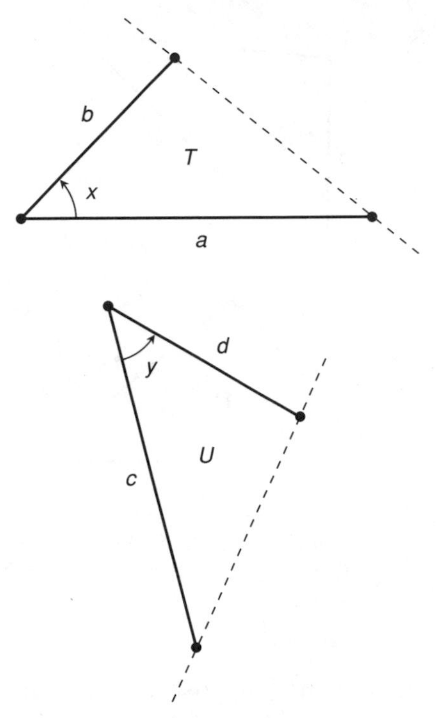

Fig. 5-10. The side-angle-side (SAS) axiom. Triangles T and U are directly congruent if and only if $a = c$, $x = y$, and $b = d$.

As an extension to this axiom, suppose T and U have two pairs of corresponding sides of equal lengths as you proceed around the triangles in opposite directions. Also, suppose the included angles between those corresponding sides have identical measures. Then T and U are inversely congruent. Also, if T and U are inversely congruent triangles, then T and U have two pairs of corresponding sides of equal lengths as you proceed around the triangles in opposite directions, and the included angles between those corresponding sides have identical measures.

THE ANGLE-SIDE-ANGLE (ASA) AXIOM

Let T and U be triangles. Suppose T and U have two pairs of corresponding angles of equal measures as you proceed around the triangles in the same direction. Also, suppose the included sides between those corresponding angles have identical lengths. Then T and U are directly congruent. Also, if T and U are directly congruent triangles, then T and U have two pairs of corresponding angles of equal measures as you proceed around the triangles in the same direction, and the included sides between those corresponding angles have identical lengths. (Fig. 5-11 can help you visualize this. In this example, $a = b$, $w = y$, and $x = z$.)

As an extension to this axiom, suppose T and U have two pairs of corresponding angles of equal measures as you proceed around the triangles in opposite directions. Also, suppose the included sides between those corresponding angles have identical lengths. Then T and U are inversely congruent. Also, if T and U are inversely congruent triangles, then T and U have two pairs of corresponding angles of equal measures as you proceed around the triangles in opposite directions, and the included sides between those corresponding angles have identical lengths.

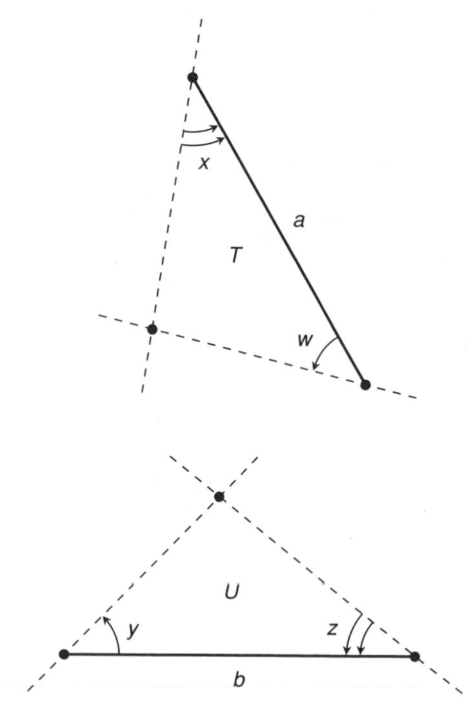

Fig. 5-11. The angle-side-angle (ASA) axiom. Triangles T and U are directly congruent if and only if $w = y$, $a = b$, and $x = z$.

THE SIDE-ANGLE-ANGLE (SAA) AXIOM

Let T and U be triangles. Suppose T and U have two pairs of corresponding angles of equal measures as you proceed around the triangles in the same direction. Suppose the corresponding sides, one of whose end points constitutes the vertex of the first-encountered angle in either triangle, have identical lengths. Then T and U are directly congruent. Also, if T and U are directly congruent triangles, then T and U have two pairs of corresponding angles of equal measures as you proceed around the triangles in the same direction, and the corresponding sides whose end points constitute the vertices of the first-encountered angles have identical lengths. (Fig. 5-12 can help you visualize this. In this example, $a = b$, $w = y$, and $x = z$.)

As an extension to this axiom, suppose T and U have two pairs of corresponding angles of equal measures as you proceed around the triangles in opposite

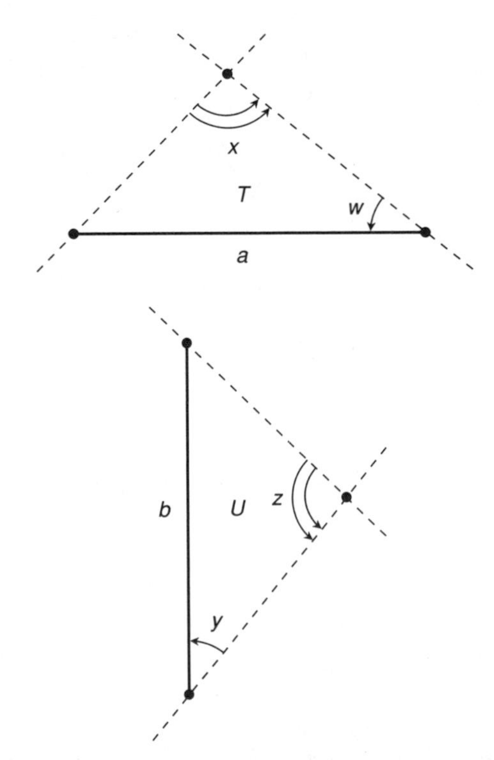

Fig. 5-12. The side-angle-angle (SAA) axiom. Triangles T and U are directly congruent if and only if $a = b$, $w = y$, and $x = z$.

directions. Suppose the corresponding sides, one of whose end points constitutes the vertex of the first-encountered angle in either triangle, have identical lengths. Then T and U are inversely congruent. Also, if T and U are inversely congruent triangles, then T and U have two pairs of corresponding angles of equal measures as you proceed around the triangles in opposite directions, and the corresponding sides whose end points constitute the vertices of the first-encountered angles have identical lengths.

Some Proofs at Last!

We're finally ready to prove a few theorems. In every case, the proofs are given in verbal form, and also as statements/reasons (S/R) tables. The theorems in this section are stated as "problems," and the proofs are stated as "solutions."

PROBLEM 5-5

Let P, Q, R, and S be mutually distinct points, all of which lie in the same plane. Suppose lines PQ, QR, RS, and SP are all mutually distinct (that is, no two of them coincide). Consider line segments PQ, QR, RS, and SP, each of which lies on the line having the same name. The line segments form a four-sided figure $PQRS$, with vertices at points P, Q, R, and S, in that order proceeding counterclockwise. Suppose the following are true:

- Line segment PQ has the same length as line segment SR
- Line segment SP has the same length as line segment RQ

Consider the line segment SQ, which divides the figure $PQRS$ into two triangles, ΔSPQ and ΔQRS. Prove that $\Delta SPQ \equiv \Delta QRS$.

SOLUTION 5-5

It helps to draw a diagram of this situation. If you wish, you can do this based on the description of the problem, and get something like Fig. 5-13.

First, let's decide upon (that is, assign) corresponding sides for the triangles, proceeding counterclockwise around the triangle in either case:

- Line segment SP in ΔSPQ corresponds to line segment QR in ΔQRS
- Line segment PQ in ΔSPQ corresponds to line segment RS in ΔQRS
- Line segment QS in ΔSPQ corresponds to line segment SQ in ΔQRS

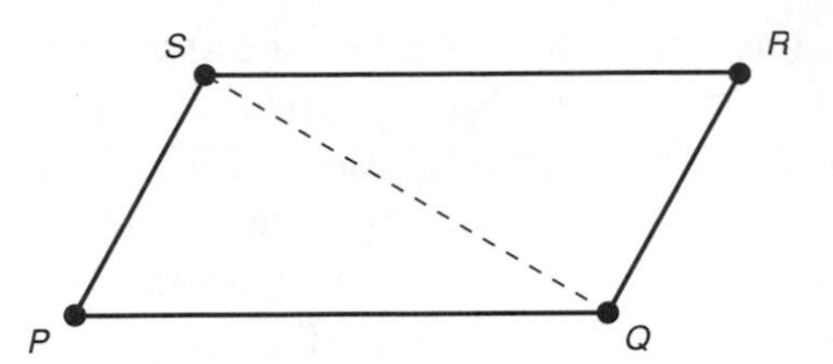

Fig. 5-13. Illustration for Problem 5-5.

We are told that line segment PQ has the same length as line segment SR. The length of a line segment does not depend on the direction in which it is expressed. Therefore, line segment PQ in ΔSPQ has the same length as line segment RS in ΔQRS. These line segments are corresponding sides in the triangles.

We are also told that line segment SP has the same length as line segment RQ. Therefore, line segment SP in ΔSPQ has the same length as line segment QR in ΔQRS. These line segments are corresponding sides in the triangles.

Finally, from the definition of the length of a line segment, we know that line segment QS in ΔSPQ has the same length as line segment SQ in ΔQRS. This line segment constitutes corresponding sides (which also happen to coincide) in the triangles.

We have shown that the corresponding sides of ΔSPQ and ΔQRS have identical lengths when we proceed around the triangles in the same direction, counterclockwise in this case. Therefore, according to the SSS axiom, we can conclude that that $\Delta SPQ \equiv \Delta QRS$. Table 5-1 is an S/R version of this proof.

PROBLEM 5-6

Let P, Q, R, and S be mutually distinct points, all of which lie in the same plane. Suppose lines PQ, QR, RS, and SP are all mutually distinct. Consider line segments PQ, QR, RS, and SP, each of which lies on the line having the same name. This forms a four-sided figure $PQRS$, with vertices at points P, Q, R, and S, in that order proceeding counterclockwise. Suppose the following are true:

- Line segment PQ has the same length as line segment SR
- Line segment SP has the same length as line segment RQ

Table 5-1. An S/R version of the proof demonstrated in Solution 5-5 and Fig. 5-13.

Statements	Reasons
Line segment *SP* in ΔSPQ corresponds to line segment *QR* in ΔQRS.	We assign them that way.
Line segment *PQ* in ΔSPQ corresponds to line segment *RS* in ΔQRS	We assign them that way.
Line segment *QS* in ΔSPQ corresponds to line segment *SQ* in ΔQRS.	We assign them that way.
Line segment *PQ* has the same length as line segment *SR*.	Given.
Line segment *SP* has the same length as line segment *RQ*.	Given.
Line segment *QS* has the same length as line segment *SQ*.	This comes from the definition of the length of a line segment: it is the same in either direction.
Corresponding sides of ΔSPQ and ΔQRS have the identical lengths expressed counterclockwise.	This is based on the above statements, and on the way we have assigned corresponding sides.
$\Delta SPQ \equiv \Delta QRS$.	SSS axiom.

Imagine lines *PQ* and *SR*, the extensions of line segments *PQ* and *SR*, respectively. Also imagine line *SQ*, the extension of line segment *SQ*. Prove that $m\angle QSR = m\angle SQP$.

SOLUTION 5-6

If you wish, you can draw a diagram of this situation based on the previous problem and on the description of this problem, and get something like Fig. 5-14.

Problem 5-5, now that it has been solved, is a theorem. In the current problem, therefore, we know that $\Delta SPQ \equiv \Delta QRS$. When two triangles are

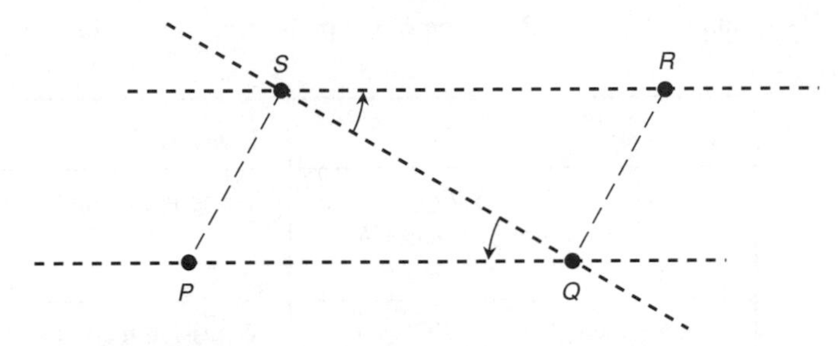

Fig. 5-14. Illustration for Problem 5-6.

directly congruent, then the counterclockwise measures of their corresponding angles, as we proceed in the same direction around either triangle, are equal. From the description of this situation (and with the help of the drawing, if you need it), it is evident that $\angle QSR$ and $\angle SQP$ are corresponding angles, the first angle in ΔQRS, and the second angle in ΔSPQ. The angle measures, as specified in the statement of this problem, are expressed counterclockwise. Therefore, their measures are equal; that is, $m\angle QSR = m\angle SQP$. Table 5-2 is an S/R version of this proof.

Table 5-2. An S/R version of the proof demonstrated in Solution 5-6 and Fig. 5-14.

Statements	Reasons
$\Delta SPQ \equiv \Delta QRS$	This is the theorem resulting from Solution 5-5.
Counterclockwise measures of corresponding angles in ΔQRS and ΔSPQ are equal.	This comes from the definition of direct congruence.
$\angle QSR$ and $\angle SQP$ are corresponding angles in ΔQRS and ΔSPQ, as we proceed in the same direction around both triangles.	This is evident from examination of the problem.
$m\angle QSR$ and $m\angle SQP$ are defined counterclockwise.	This is evident from the statement of the problem.
$m\angle QSR = m\angle SQP$	This comes from information derived in the preceding steps, and from the definition of direct congruence.

PROBLEM 5-7

In the situation of the previous problem, let T be a point on line PQ such that point Q is between point P and point T. Let U be a point on line SR such that point S is between point U and point R. Suppose that, in addition to the other conditions in the previous problem, the following are true:

- Line segment QT has the same length as line segment US
- Line segment UQ has the same length as line segment ST

Prove that $m\angle USQ = m\angle TQS$.

SOLUTION/EXERCISE 5-7

Try this for yourself! Here are some hints:

- Construct triangles ΔUQS and ΔTSQ
- Show that ΔUQS and ΔTSQ are directly congruent
- Use the same approach as in Solution 5-6
- Feel free to use Fig. 5-15 as a visual aid

If you have trouble proving this, accept it on faith for now, and come back to it tomorrow. We have all had the experience of solving a tough problem that turned out to be easy after we "slept on it." This technique often works well with elusive math proofs.

ALTERNATE INTERIOR ANGLES

Here's a new term that needs to be defined. Suppose line PQ and line SR are distinct lines that are both crossed by a transversal line SQ. Let T be a point on line PQ such that point Q is between point P and point T. Let U be a point on line

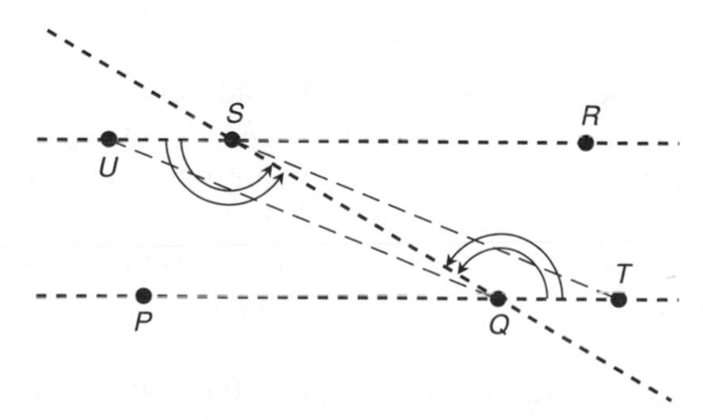

Fig. 5-15. Illustration for Problem 5-7.

SR such that point *S* is between point *U* and point *R*. Then ∠*QSR* and ∠*SQP* are a pair of *alternate interior angles* formed by line *PQ*, line *SR*, and the transversal line *SQ*. Also, ∠*USQ* and ∠*TQS* are a pair of alternate interior angles formed by line *PQ* and line *SR*, and the transversal line *SQ*. These pairs of angles are marked in Figs. 5-14 and 5-15.

PROBLEM 5-8

Prove that when two parallel lines *L* and *M* are crossed by a transversal line *T*, pairs of alternate interior angles always have equal measure. Let's call this the *AIA Theorem*.

SOLUTION/EXERCISE 5-8

Try this for yourself! Here's a hint: It follows in a straightforward way from the "combined forces" of Solution 5-6 and Solution/Exercise 5-7. Both of these, having been proven, are theorems that can be used as lemmas here.

PROBLEM 5-9

Prove that the sum of the measures of the three interior angles of a triangle is always equal to the measure of a straight angle.

SOLUTION 5-9

Consider an arbitrary triangle Δ*PQR*, with vertices at points *P*, *Q*, and *R*, expressed counterclockwise in that order. Let x be the measure of ∠*RQP*; let y be the measure of ∠*PRQ*; let z be the measure of ∠*QPR*. Fig. 5-16A illustrates an example of such a triangle.

Choose a point *S* somewhere outside Δ*PQR*, such that line *RS* is parallel to line *PQ*. Then choose a point *T* on line *RS*, such that point *R* is between point *T* and point *S*. This gives us a situation in which there are two parallel lines, crossed by two different transversals. The parallel lines are line *PQ* and line *RS*. The transversals are line *PR* and line *QR*, as shown in Fig. 5-16B.

Now consider ∠*QRS*, and call its measure x^*. Also consider ∠*TRP*, and call its measure z^*. Notice that ∠*RQP* and ∠*QRS*, whose measures are x and x^* respectively, are alternate interior angles, defined by the transversal line *QR*. Also notice that ∠*QPR* and ∠*TRP*, whose measures are z and z^* respectively, are alternate interior angles, defined by the transversal line *PR*. According to the AIA Theorem, it follows that $x = x^*$, and also that $z = z^*$.

It is evident from the geometry of the situation that the sum $z^* + y + x^*$ adds up to an angle whose measure is a straight angle. (This is also

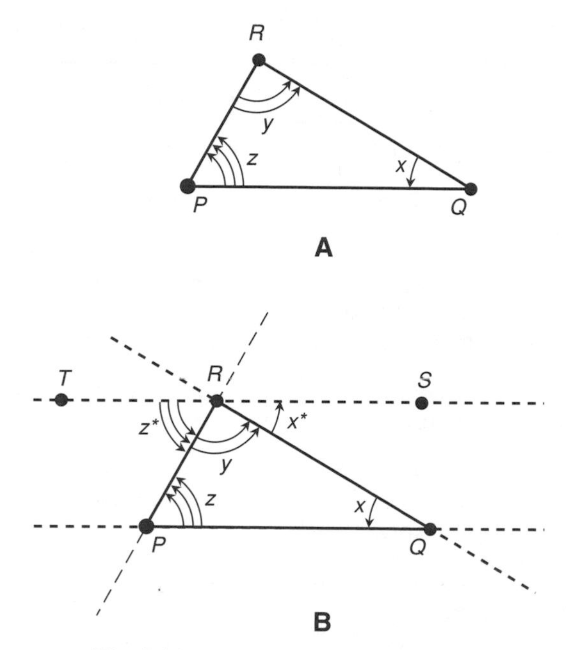

Fig. 5-16. Illustration for Problem 5-9.

called a *180° angle*, based on the angular degree, symbolized °, which represents 1/360 of a full circle.) In other words, $z^* + y + x^* = 180°$. The straight angle in this case is $\angle TRS$, formed by the collinear points T, R, and S. It follows that that $z + y + x = 180°$. (You'll get a chance to provide the reason in Quiz Question 1.) Notice that x, y, and z are the measures of the interior angles of ΔPQR. Therefore, the sum of the measures of the interior angles of ΔPQR is equal to the measure of a straight angle. Table 5-3 is an S/R version of this proof.

PROBLEM 5-10

Suppose ΔPQR is an isosceles triangle defined by three distinct points P, Q, and R, expressed counterclockwise in that order. Let p be the length of line segment QR. Let q be the length of line segment RP. Let r be the length of line segment PQ. Suppose $q = r$. Let S be the point at the center of line segment QR, such that line segment QS has the same length as line segment RS. Line segment PS divides ΔPQR into two triangles, ΔPQS and ΔPSR. Prove that ΔPQS and ΔPSR are inversely congruent.

SOLUTION 5-10

The trick here is to work around the two triangles, ΔPQS and ΔPSR, in opposite directions. For that reason, let's rename the second triangle

Table 5-3. An S/R version of the proof demonstrated in Solution 5-9 and Fig. 5-16.

Statements	**Reasons**
Consider a triangle ΔPQR.	We have to start somewhere.
Choose a point S such that line RS is parallel to line PQ.	We will use this point later.
Choose a point T on line RS, such that point R is between point T and point S.	We will use this point later.
Line PR and line QR are transversals to the parallel lines PQ and RS.	This is apparent from the definition of a transversal line.
Consider $\angle QRS$, and call its measure x^*. Also consider $\angle TRP$, and call its measure z^*.	We will use these later.
Angles $\angle RQP$ and $\angle QRS$ are alternate interior angles.	This is apparent from the definition of alternate interior angles.
Angles $\angle QPR$ and $\angle TRP$ are alternate interior angles.	This is apparent from the definition of alternate interior angles.
$x = x^*$	This follows from the AIA Theorem.
$z = z^*$	This follows from the AIA Theorem.
$z^* + y + x^* = 180°$	This is apparent from the geometry of the situation, and from the definition of an angular degree.
$z + y + x = 180°$	You'll get a chance to fill this in later.
The measures of the interior angles of ΔPQR add up to measure of a straight angle.	This is because x, y, and z are the measures of the interior angles of ΔPQR.

ΔPRS. That way, we go counterclockwise around ΔPQS, and clockwise around ΔPRS.

Let q^* be the length of line segment PS. Let p^* be the length of line segment QS. Let p^{**} be the length of line segment RS. If you wish, you can draw a diagram of this situation. The result should look something like Fig. 5-17. We are told that $q = r$. We know that $p^* = p^{**}$. (You'll

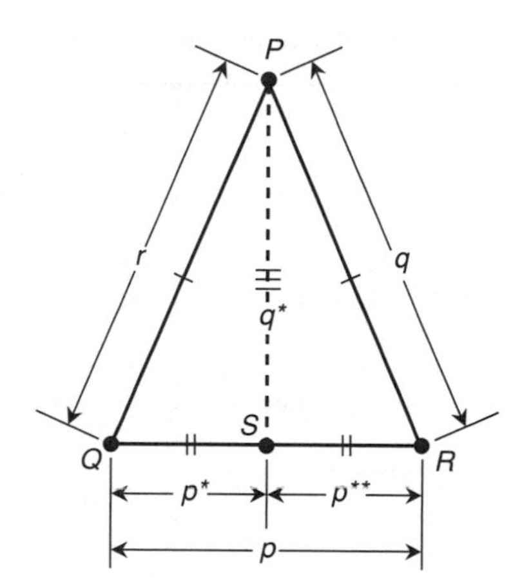

Fig. 5-17. Illustration for Problem 5-10.

get a chance to provide the reason in Quiz Question 2 at the end of this chapter.) We also know that $q^* = q^*$; this is trivial! Let's state these three equations in order:

$$q = r$$
$$p^* = p^{**}$$
$$q^* = q^*$$

If we proceed counterclockwise around ΔPQS, we encounter sides of lengths r, p^*, and q^* in that order. If we proceed clockwise around ΔPRS, we encounter sides of lengths q, p^{**}, and q^* in that order. In this second case, we can substitute r for q because $q = r$, we can substitute p^* for p^{**} because $p^* = p^{**}$, and we know $q^* = q^*$. This means that if we proceed clockwise around ΔPRS, we encounter sides of lengths r, p^*, and q^* in that order. These are the same lengths, in the same order, as the lengths of the sides we encounter when we go counterclockwise around ΔPQS. Therefore, by definition, the two triangles are inversely congruent. Table 5-4 is an S/R version of this proof.

PROBLEM 5-11

Imagine a *regular hexagon*. This is a geometric figure with six vertices, six straight sides of identical length that connect adjacent pairs of vertices, and six interior angles of identical measure. As we go counterclockwise around the hexagon, let's name the vertices P, Q, R, S, T, and U in that order. Let m be the length of each side. Let x be the measure of each interior angle. Prove that $\Delta PQR \equiv \Delta STU$.

Table 5-4. An S/R version of the proof demonstrated in Solution 5-10 and Fig. 5-17.

Statements	**Reasons**
Let q^* be the length of line segment *PS*.	We need to call it something!
Let p^* be the length of line segment *QS*.	We need to call it something!
Let p^{**} be the length of line segment *RS*.	We need to call it something!
$q = r$	We are told this.
$p^* = p^{**}$	You'll get a chance to fill this in later.
$q^* = q^*$	This is trivial. Anything is equal to itself.
Counterclockwise around ΔPQS, we encounter sides of lengths r, p^*, and q^* in that order.	This is evident from the geometry of the situation.
Clockwise around ΔPRS, we encounter sides of lengths q, p^{**}, and q^* in that order.	This is evident from the geometry of the situation.
Clockwise around ΔPRS, we encounter sides of lengths r, p^*, and q^* in that order.	This follows from substituting r for q and p^* for p^{**} in the preceding statement.
$\Delta PQS \equiv\text{-}\ \Delta PSR$.	This follows from the statements in the first and third lines above this line, and from the definition of inverse congruence for triangles.

SOLUTION 5-11

If you wish, you can draw a diagram. The result should look something like Fig. 5-18. In ΔPQR, consider line segment *PQ* and line segment *QR*. Both of these sides have length m because they are sides of the hexagon, and we are told that all the sides of the hexagon have length m. Consider the included angle between these two sides. It has measure x because it is an interior angle of the hexagon, and we are told that all the interior angles of the hexagon have measure x. Proceeding counterclockwise from point *P* around ΔPQR, we encounter a side of length m, then an angle of measure x, and then a side of length m.

In ΔSTU, consider line segment *ST* and line segment *TU*. Both of these sides have length m because they are sides of the hexagon, and we

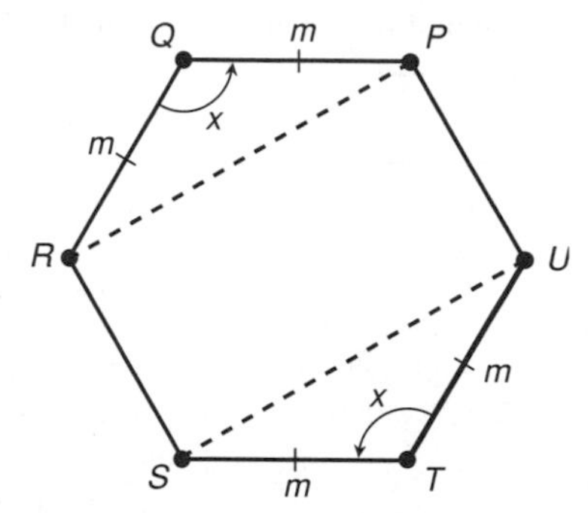

Fig. 5-18. Illustration for Problem 5-11.

are told that all the sides of the hexagon have length m. Consider the included angle between these two sides. It has measure x because it is an interior angle of the hexagon, and we are told that all the interior angles of the hexagon have measure x. Proceeding counterclockwise from point S around ΔSTU, we encounter a side of length m, then an angle of measure x, and then a side of length m. These side lengths and included angle measure are the same, and occur in the same order, as the corresponding side lengths and included angle measure in ΔPQR. Therefore, it follows that $\Delta PQR \equiv \Delta STU$. (You'll get a chance to provide the reason in Quiz Question 3.) Table 5-5 is an S/R version of this proof.

PROBLEM 5-12

Let P, Q, R, and S be mutually distinct points, all of which lie in the same plane. Suppose lines PQ and RS are parallel. Suppose also that line segment PQ has the same length as line segment RS. Imagine two transversal lines SQ and PR, both of which cross lines PQ and RS, and which intersect at a point T. Prove that $\Delta PQT \equiv \Delta RST$.

SOLUTION 5-12

If you wish, you can draw a diagram. The result should look something like Fig. 5-19. Transversal line PR crosses both parallel lines PQ and RS. There is a pair of alternate interior angles formed by these three lines: $\angle SRT$ and $\angle QPT$. These two angles have equal measure. (You'll get a chance to provide the reason in Quiz Question 6.) Let x be this measure.

Let m be the length of line segments PQ and RS. We can use m for both, because we are told their lengths are equal.

Transversal line SQ crosses both parallel lines PQ and RS. There is a pair of alternate interior angles formed by these three lines. These angles are $\angle TQP$ and $\angle TSR$. These two angles have equal measure.

Table 5-5. An S/R version of the proof demonstrated in Solution 5-11 and Fig. 5-18.

Statements	Reasons
In ΔPQR, line segments PQ and QR both have length m.	Line segments PQ and QR are sides of of the hexagon, and we are told that all sides of the hexagon have length m.
In ΔPQR, the included angle between the adjacent sides of length m has measure x.	This included angle is an interior angle of the hexagon, and we are told that all interior angles of the hexagon have measure x.
Counterclockwise from point P around ΔPQR, we encounter a side of length m, then an angle of measure x, and finally a side of length m.	This is evident from the geometry of the situation.
In ΔSTU, line segments ST and TU both have length m.	Line segments ST and TU are sides of the hexagon, and we are told that all sides of the hexagon have length m.
In ΔSTU, the included angle between the adjacent sides of length m has measure x.	This included angle is an interior angle of the hexagon, and we are told that all interior angles of the hexagon have measure x.
Counterclockwise from point S around ΔSTU, we encounter a side of length m, then an angle of measure x, and finally a side of length m.	This is evident from the geometry of the situation.
ΔPQR and ΔSTU have pairs of corresponding sides of identical lengths, with included angles of identical measures.	This is evident from the geometry of the situation.
$\Delta PQR \equiv \Delta STU$.	You'll get a chance to fill this in later.

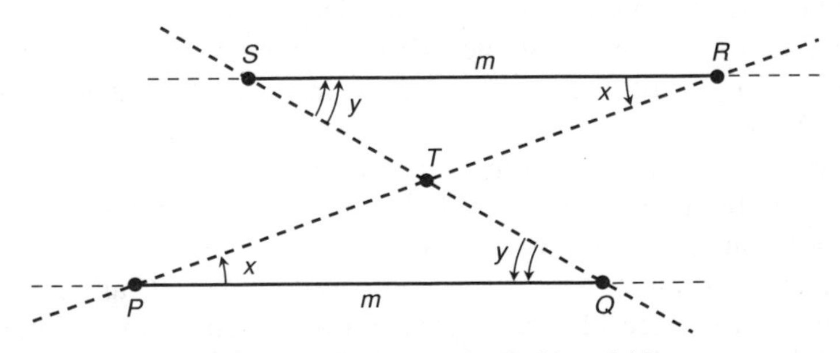

Fig. 5-19. Illustration for Problem 5-12.

Table 5-6. An S/R version of the proof demonstrated in Solution 5-12 and Fig. 5-19.

Statements	Reasons
Transversal line *PR* crosses lines *PQ* and *RS*.	This is evident from the geometry of of the situation.
The two angles $\angle SRT$ and $\angle QPT$ are alternate interior angles within parallel lines.	This is evident from the geometry of the situation.
$\angle SRT$ and $\angle QPT$ have equal measure.	You'll get a chance to fill this in later.
Let x be the measure of $\angle SRT$ and $\angle QPT$.	We have to call it something!
Let m be the length of line segments *PQ* and *RS*.	We're told that their lengths are equal, and we have to call them something!
Transversal line *SQ* crosses lines *PQ* and *RS*.	This is evident from the geometry of the situation.
The two angles $\angle TQP$ and $\angle TSR$ are alternate interior angles within parallel lines.	This is evident from the geometry of the situation.
$\angle TQP$ and $\angle TSR$ have equal measure.	You'll get a chance to fill this in later.
Let y be the measure of $\angle TQP$ and $\angle TSR$.	We have to call it something!
Counterclockwise around ΔPQT or ΔRST, we encounter an angle of measure x, a side of length m, and an angle of measure y, in that order.	This is evident from the geometry of the situation.
$\Delta PQT \equiv \Delta RST$.	This follows from the ASA axiom.

(You'll get a chance to provide the reason in Quiz Question 6.) Let y be this measure.

Proceeding from point P counterclockwise around ΔPQT, we encounter first an angle of measure x, then a side of length m, and then an angle of measure y. Proceeding from point R counterclockwise around ΔRST, we encounter first an angle of measure x, then a side of length m, and then an angle of measure y. According to the ASA axiom, therefore, $\Delta PQT \equiv \Delta RST$. Table 5-6 is an S/R version of this proof.

PROBLEM 5-13

Let P, Q, R, and S be mutually distinct points, all of which lie in the same plane. Suppose lines *PQ* and *RS* are parallel. Suppose also that

$\angle PSR$ and $\angle RQP$ are right angles. Imagine a transversal line *PR* that crosses lines *PQ* and *RS*. Prove that $\Delta RPQ \equiv \Delta PRS$.

SOLUTION 5-13

If you wish, you can draw a diagram. The result should look something like Fig. 5-20. Let *m* be the length of line segment *RP*, which is the same as line segment *PR*.

A pair of alternate interior angles is formed by the transversal line *PR*, line *PQ*, and line *RS*. These angles are $\angle SRP$ and $\angle QPR$. According to the AIA Theorem, they have equal measure. Let *x* be this measure.

We are told that $\angle PSR$ and $\angle RQP$ are right angles. Therefore, they have equal measure. (You'll get a chance to provide the reason in Quiz Question 8.) Let *y* be this measure.

Table 5-7. An S/R version of the proof demonstrated in Solution 5-13 and Fig. 5-20.

Statements	Reasons
The length of line segment *RP* equals the length of line segment *PR*.	It doesn't matter which direction we go when we define the length of a line segment.
Let *m* be the length of line segments *RP* and *PR*.	We have to call it something!
$\angle SRP$ and $\angle QPR$ are alternate interior angles.	This is apparent from the geometry of the situation.
$m\angle SRP = m\angle QPR$.	This follows from the AIA Theorem.
Let *x* be the measure of $\angle SRP$ and $\angle QPR$.	We have to call it something!
$\angle PSR$ and $\angle RQP$ are right angles.	We are told this.
$m\angle PSR = m\angle RQP$.	You'll get a chance to fill this in later.
Let *y* be the measure of $\angle PSR$ and $\angle RQP$.	We have to call it something!
Counterclockwise around ΔRPQ or ΔPRS, we encounter a side of length *m*, an angle of measure *x*, and an angle of measure *y*, in that order.	This is evident from the geometry of the situation.
$\Delta RPQ \equiv \Delta PRS$.	This follows from the SAA axiom.

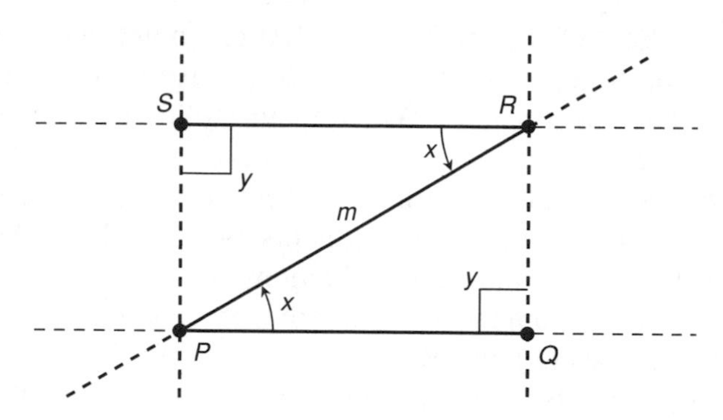

Fig. 5-20. Illustration for Problem 5-13.

Proceeding from point R counterclockwise around ΔRPQ, we encounter first a side of length m, then an angle of measure x, and then an angle of measure y. Proceeding from point P counterclockwise around ΔPRS, we encounter first a side of length m, then an angle of measure x, and then an angle of measure y. According to the SAA axiom, therefore, $\Delta RPQ \equiv \Delta PRS$. Table 5-7 is an S/R version of this proof.

Quiz

This is an "open book" quiz. You may refer to the text in this chapter. A good score is 8 correct. Answers are in the back of the book.

1. Refer to Solution 5-9 and Table 5-3. At one point in this proof, you are told that you'll get a chance to provide a reason for the statement later. You now have that chance! Which of the following sentences provides the correct reason?
   (a) This follows from the SAA axiom.
   (b) We can substitute z for z^* and x for x^* in the preceding equation, because $z = z^*$ and $x = x^*$.
   (c) This follows from the definition of a triangle.
   (d) This follows from the parallel axiom.

2. Refer to Solution 5-10 and Table 5-4. At one point in this proof, you are told that you'll get a chance to provide a reason for the statement later. You now have that chance! Which of the following sentences provides the correct reason?
   (a) These are the lengths of line segments *QS* and *RS*, respectively, and we are told that these line segments have equal lengths.
   (b) This follows from the parallel axiom.
   (c) This follows from the definition of a perpendicular lines; in this case they are lines *QR* and *PS*.
   (d) They have to be equal. If they were not, line *PS* would not intersect line *QR*, but these two lines obviously do intersect.

3. Refer to Solution 5-11 and Table 5-5. At one point in this proof, you are told that you'll get a chance to provide a reason for the statement later. You now have that chance! Which of the following sentences provides the correct reason?
   (a) This follows from the ASA axiom.
   (b) This follows from the AAA axiom.
   (c) This follows from the SAS axiom.
   (d) This follows from the SAA axiom.

4. Unless otherwise specified, an angle is always expressed
   (a) clockwise.
   (b) counterclockwise.
   (c) from left to right.
   (d) from right to left.

5. Suppose we are told that lines *L*, *M*, and *N* are mutually distinct. Which, if any, of the following scenarios (a), (b), or (c) is impossible?
   (a) Line *L* is perpendicular to line *M*, and line *L* is also perpendicular to line *N*.
   (b) Line *M* is perpendicular to line *L*, and line *M* is also perpendicular to line *N*.
   (c) Line *N* is perpendicular to line *L*, and line *N* is also perpendicular to line *M*.
   (d) All three scenarios (a), (b), and (c) are possible.

6. Refer to Solution 5-12 and Table 5-6. At two points in this proof, you are told that you'll get a chance to provide a reason for the state-

ment later. You now have that chance! The same sentence can be inserted in both places to complete the proof. Which of the following sentences is it?

(a) This follows from the ASA axiom, and the fact that line *PQ* is parallel to line *RS*.
(b) This follows from the SAA axiom, and the fact that line *PQ* is parallel to line *RS*.
(c) This follows from the definition of parallel lines, which tells us that line *PQ* is parallel to line *RS*.
(d) This follows from the AIA Theorem, and the fact that line *PQ* is parallel to line *RS*.

7. What is the distinction between "direct" and "inverse" as these words apply to congruent triangles?
   (a) When two triangles are directly congruent, is it never necessary to rotate one of them in order to get it to coincide with the other, but when two triangles are inversely congruent, it is usually necessary to rotate one of them in order to get it to coincide with the other.
   (b) When two triangles are directly congruent, is it usually necessary to rotate one of them in order to get it to coincide with the other, but when two triangles are inversely congruent, it is never necessary to rotate one of them in order to get it to coincide with the other.
   (c) When two triangles are directly congruent, is it never necessary to flip one of them over in order to get it to coincide with the other, but when two triangles are inversely congruent, it is usually necessary to flip one of them over in order to get it to coincide with the other.
   (d) When two triangles are directly congruent, is it usually necessary to flip one of them over in order to get it to coincide with the other, but when two triangles are inversely congruent, it is never necessary to flip one of them over in order to get it to coincide with the other.

8. Refer to Solution 5-13 and Table 5-7. At one point in this proof, you are told that you'll get a chance to provide a reason for the statement later. You now have that chance! Which of the following sentences provides the correct reason?
   (a) This follows from the SAA axiom.
   (b) This follows from the right angle axiom.
   (c) This follows from the two-point axiom.
   (d) This follows from the parallel axiom.

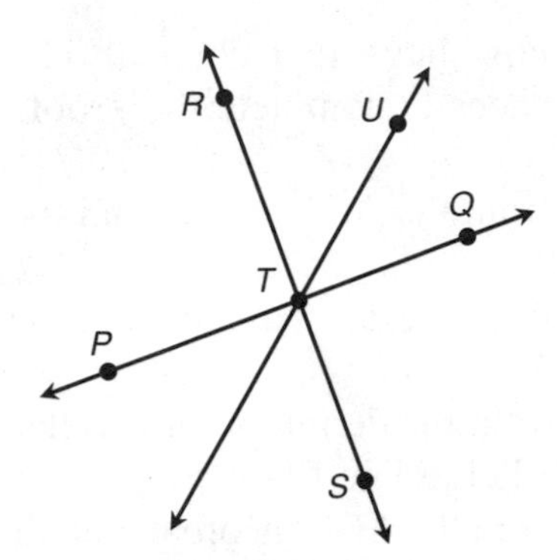

Fig. 5-21. Illustration for Quiz Questions 9 and 10.

9. In Fig. 5-21, suppose all three lines intersect at a common point *T*. Suppose we are told that $\angle QTR$ is a right angle. It follows that line *PQ* is perpendicular to line *RS* from
   (a) the two-point axiom.
   (b) the right-angle axiom.
   (c) the extension axiom.
   (d) the definition of perpendicular lines.

10. In Fig. 5-21, suppose all three lines intersect at a common point *T*. Suppose we are told that $\angle QTR$ is a right angle. It follows that $\angle QTU$ and $\angle UTR$
    (a) have equal measure.
    (b) are complementary.
    (c) are supplementary.
    (d) are alternate interior angles.

Sets and Numbers

This chapter will familiarize you with some of the ways accepted facts can be logically combined to prove theorems about sets and numbers. Get ready for another mind drill!

Some Definitions

Again, let's start with definitions. It's worth repeating: A good definition must stand on its own. Drawings can help us visualize or conceptualize definitions, but they should never be a necessary part of a definition.

SETS AND ELEMENTS

A *set* is a collection or group of objects called *elements* or *members*. Set elements can be almost anything, such as:

- Points
- Objects
- People
- Numbers
- Cities
- Planets
- Stars
- Galaxies

If an object a is an element of set A, then this fact is written as:

$$a \in A$$

If an object b is not an element of set A, then this fact is written as:

$$b \notin A$$

LISTING OF SET ELEMENTS

When the elements of a set can be listed, the elements are written one after another, separated by commas, and enclosed in curly brackets. An example of a set with five elements is:

$$S = \{2, 4, 6, 8, 10\}$$

CONSTANT

A *constant* is a specific element of a set. Constants are often, but not always, denoted by lowercase, italicized letters from the first part of the alphabet, such as a, b, and c.

VARIABLE

A *variable* can represent any element in a given set. Variables are often, but not always, denoted by lowercase, italicized letters from the last part of the alphabet, such as x, y, and z.

SET INTERSECTION

The *intersection* of two sets A and B, written $A \cap B$, is the set of elements that are in both set A and set B. It is a set C such that for every element x:

$$x \in C \Leftrightarrow (x \in A) \,\&\, (x \in B)$$

SET UNION

The *union* of two sets A and B, written $A \cup B$, is the set of elements that are in set A, set B, or both. It is a set D such that for every element x:

$$x \in D \Leftrightarrow (x \in A) \vee (x \in B)$$

COINCIDENT SETS

Two non-empty sets A and B are *coincident*, written $A = B$, if and only if they both contain exactly the same elements. That is, for every element x:

$$x \in A \Leftrightarrow x \in B$$

SUBSET

Set A is a *subset* of set B, written $A \subseteq B$, if and only if every element x in set A is also in set B. That is, for every element x:

$$x \in A \Rightarrow x \in B$$

PROPER SUBSET

Set A is a *proper subset* of set B, written $A \subset B$, if and only if every element x in set A is also in set B, but the two sets are not coincident. That is, for every element x:

$$x \in A \Rightarrow x \in B$$
$$A \neq B$$

EMPTY SET

The *empty set*, also called the *null set*, is the set containing no elements. It is denoted either by the "circle-slash" symbol $\varnothing$, or by a pair of curly brackets $\{ \}$ with a blank space in between.

DISJOINT SETS

Two non-empty sets A and B are *disjoint* if and only if they have no elements in common; that is, if and only if their intersection is the null set.

CARDINALITY

The *cardinality* of a set A is the number of distinct elements a such that $a \in A$.

VENN DIAGRAMS

Union and intersection of sets can be illustrated by *Venn diagrams*. Fig. 6-1 is a Venn diagram that shows the intersection of two sets that are non-disjoint (they overlap) and non-coincident (they are not identical). Set $A \cap B$ is the cross-hatched area, common to both sets A and B. Fig. 6-2 shows the union of the same two sets. Set $A \cup B$ is the shaded area, representing elements that are in set A or in set B, or both.

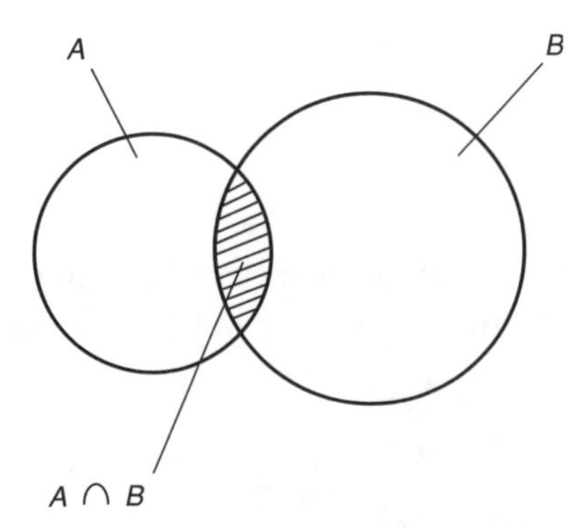

Fig. 6-1. Intersection of non-disjoint, non-coincident sets A and B.

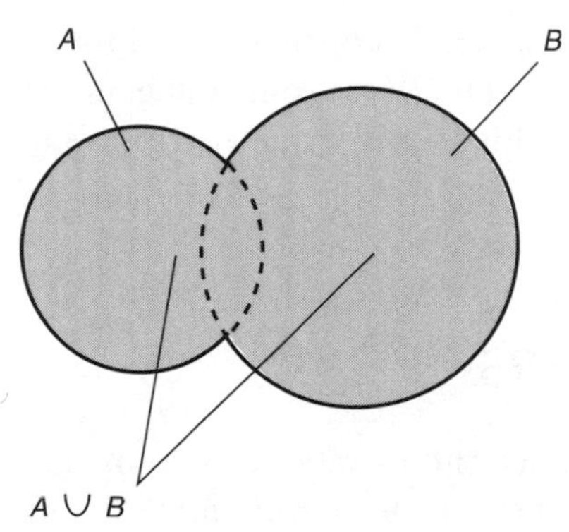

Fig. 6-2. Union of nondisjoint, noncoincident sets A and B.

NUMBER

A *number* is an abstract, unambiguous expression of a mathematical quantity.

NATURAL NUMBERS

Natural numbers, also called *whole numbers* or *counting numbers*, are built up from a starting point of 0 or 1, depending on which text you consult. The set of natural numbers is denoted $\boldsymbol{N}$. Usually we include 0, obtaining:

$$\boldsymbol{N} = \{0, 1, 2, 3, \ldots, n, \ldots\}$$

In some texts, 0 is not included, so:

$$\boldsymbol{N} = \{1, 2, 3, 4, \ldots, n, \ldots\}$$

INTEGERS

A "mirror-image" of the set of natural numbers is obtained when each element of $\boldsymbol{N}$ is multiplied by -1:

$$-\boldsymbol{N} = \{0, -1, -2, -3, \ldots\}$$

The union of this set with the set of natural numbers (with 0 included) produces the set of *integers*, commonly denoted $\boldsymbol{Z}$:

$$\begin{aligned} \boldsymbol{Z} &= \boldsymbol{N} \cup -\boldsymbol{N} \\ &= \{\ldots, -3, -2, -1, 0, 1, 2, 3, \ldots\} \end{aligned}$$

The set of natural numbers is a proper subset of the set of integers. Every natural number is an integer, but there are some integers (–1, –2, –3, and so on) that are not natural numbers. This fact is written as follows:

$$\boldsymbol{N} \subset \boldsymbol{Z}$$

RATIONAL NUMBERS

A *rational number* is a quotient, or ratio, of two integers, where the denominator is not zero. In this context, the term *rational* derives from the word *ratio*. The standard form for a rational number r is:

$$r = a/b$$

where a and b are integers, and $b \neq 0$. The set of all possible such quotients encompasses the entire set of rational numbers, denoted $\boldsymbol{Q}$. Thus,

$$\boldsymbol{Q} = \{r \mid r = a/b, a \in \boldsymbol{Z}, b \in \boldsymbol{Z}, b \neq 0\}$$

The vertical line in the above expression translates as "such that." Sometimes a colon is used instead. In that case we write:

$$\boldsymbol{Q} = \{r : r = a/b, a \in \boldsymbol{Z}, b \in \boldsymbol{Z}, b \neq 0\}$$

In some texts (and in Chapter 3 of this book), the value of b is restricted to the set of positive integers. The definition here is equivalent to the one in Chapter 3. Would you like to prove it as an exercise?

The set of integers is a proper subset of the set of rational numbers. Every integer is a rational number, but there are some rational numbers that are not integers. We write:

$$\boldsymbol{Z} \subset \boldsymbol{Q}$$

IRRATIONAL NUMBERS

An *irrational number* is a number that lies between some pair of different integers, but that *cannot* be expressed as the ratio of any two integers. This means that if k is an irrational number, then there exist *no* two integers a and b such that $k = a/b$. Examples of irrational numbers include:

- The length of the diagonal of a square measuring 1 unit on each edge
- The ratio of the circumference to the diameter of a circle in a plane

The set of irrational numbers can be denoted $\boldsymbol{S}$. No irrational number is rational, and no rational number is irrational:

$$\boldsymbol{S} \cap \boldsymbol{Q} = \varnothing$$

REAL NUMBERS

The set of *real numbers*, denoted $\boldsymbol{R}$, is the union of the sets of rational and irrational numbers:

$$\boldsymbol{R} = \boldsymbol{Q} \cup \boldsymbol{S}$$

The real numbers are related to the rational numbers, the integers, and the natural numbers as follows:

$$\boldsymbol{N} \subset \boldsymbol{Z} \subset \boldsymbol{Q} \subset \boldsymbol{R}$$

ELEMENTARY TERMS

Let's consider the following terms elementary. You should have a good idea of what they mean and how they behave, having worked with them since you were in grade school.

- Two constants or variables are *equal* if and only if they have the same value. The symbol for *equality* is a pair of parallel double dashes (=). Two constants or variables are *unequal* or *not equal* if and only if they have different values. The symbol for *inequality* is an equals sign with a forward slash through it (≠).
- The operation of *addition* is applicable to two or more numbers. The arguments are called *addends*, and the resultant is called the *sum*. The addition symbol is a symmetrical, upright cross (+).
- The operation of *subtraction* is applicable to two numbers. The first argument is called the *minuend*, the second argument is called the *subtrahend*, and the resultant is called the *difference*. The subtraction symbol is a long dash (−).
- The operation of *multiplication* is applicable to two or more numbers. The arguments are called *factors*, and the resultant is called the *product*. The multiplication symbol can be a slanted cross (×), an elevated dot (·), or, for constants and variables represented by letters of the alphabet, nothing at all. For example, xy represents the product of x and y.

- The operation of *division*, also called the *ratio operation*, is applicable to two numbers. The first argument is called the *numerator*, the second argument is called the *denominator*, and the resultant is called the *quotient* or the *ratio*. The division symbol can be a double dash with two dots (÷), a forward slash (/), or a colon (:).

Axioms

Let's accept the following rules of arithmetic without proof. We already have the rules of logic, and we already have the definitions we need. Here are the assumptions!

EQUALITY AXIOM

For all real numbers r, s, and t, the following equations hold:

$$r = r$$
$$(r = s) \Rightarrow (s = r)$$
$$[(r = s) \;\&\; (s = t)] \Rightarrow (r = t)$$

The first of these facts is known as the *reflexive property*. The second statement is called the *symmetric property*. The third statement is called the *transitive property*. Because equality obeys all three of these rules, equality is said to be an *equivalence relation*.

SUM-OF-INTEGERS AXIOM

For any two integers a and b, the sum $a + b$ is an integer.

PRODUCT-OF-INTEGERS AXIOM

For any two integers a and b, the product ab is an integer. Also, for any two integers a and b, the following statement holds:

$$[(a = 0) \vee (b = 0)] \Leftrightarrow (ab = 0)$$

ADDITIVE-INVERSE AXIOM

For any real number r, there exists a unique real number $-r$ called the *additive inverse* of r, such that both of the following equations hold:

$$-r = -1 \times r$$
$$r + (-r) = 0$$

SUM-OF-FRACTIONS AXIOM

For any four integers a, b, c, and d where $b \neq 0$ and $d \neq 0$, the following equation holds:

$$a/b + c/d = (ad + bc)/bd$$

DIFFERENCE-BETWEEN-FRACTIONS AXIOM

For any four integers a, b, c, and d where $b \neq 0$ and $d \neq 0$, the following equation holds:

$$a/b - c/d = a/b + (-c/d)$$

PRODUCT-OF-FRACTIONS AXIOM

For any four integers a, b, c, and d where $b \neq 0$ and $d \neq 0$, the following equation holds:

$$(a/b)(c/d) = (ac)/(bd)$$

QUOTIENT-OF-FRACTIONS AXIOM

For any four integers a, b, c, and d where $b \neq 0$, $c \neq 0$, and $d \neq 0$, the following equation holds:

$$(a/b)/(c/d) = (ad)/(bc)$$

DIVISION-BY-1 AXIOM

For any real number r, the following equation holds:

$$r/1 = r$$

COMMUTATIVE AXIOM FOR ADDITION

For any two real numbers r and s, the following equation holds:

$$r + s = s + r$$

COMMUTATIVE AXIOM FOR MULTIPLICATION

For any two real numbers r and s, the following equation holds:

$$rs = sr$$

DISTRIBUTIVE AXIOM

For any three real numbers r, s, and t, the following equation holds:

$$r(s + t) = rs + rt$$

Some Proofs at Last!

We are now ready to prove some theorems about sets and numbers. Proofs are given in verbal form, and also as S/R tables. In some of these proofs, it may seem as if we're "going around the world to get uptown." But remember: When we want to prove something rigorously, we must be painstaking in the execution.

PROBLEM 6-1

Prove that any element in the intersection of two sets is also in their union.

SOLUTION 6-1

Let A and B be non-empty sets. Let c be a constant. We must prove that if $c \in A \cap B$, then $c \in A \cup B$. If you wish, you can draw a Venn dia gram to illustrate this scenario. You should get something that looks like Fig. 6-3. The cross-hatched region represents $A \cap B$, and the shaded region represents $A \cup B$. Note the location of the point representing c.

We are told that $c \in A \cap B$, that is, c is an element of the intersection of sets A and B. From the definition of set intersection, c is an ele-

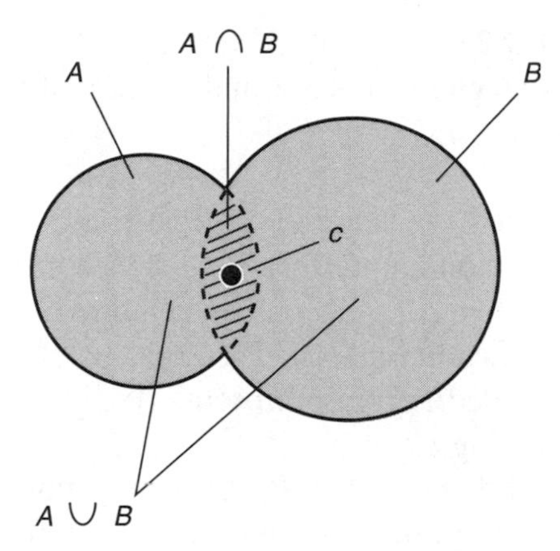

Fig. 6-3. Illustration for Problem 6-1.

ment of set A and is also an element of set B. From the definition of logical conjunction given in Chapter 1, $c \in A$. From the definition of logical disjunction given in Chapter 1, c is an element of set A or set B. From the definition of set union, it follows that c is an element of the union of sets A and B, that is, $c \in A \cup B$. Table 6-1 is an S/R version of this proof.

Table 6-1. An S/R version of the proof demonstrated in Solution 6-1 and Fig. 6-3.

Statements	**Reasons**
Let A and B be non-empty sets.	We will use these in the proof.
Let c be a constant.	We will use this in the proof.
Assume $c \in A \cap B$.	This is our initial assumption.
$(c \in A)$ & $(c \in B)$.	This follows from the definition of set intersection.
$c \in A$.	This follows from the definition of logical conjunction.
$(c \in A) \vee (c \in B)$.	This follows from the definition of logical disjunction.
$c \in A \cup B$.	This follows from the definition of set union.

PROBLEM 6-2

Prove that if an element is not in the union of two sets, then it is not in their intersection.

SOLUTION 6-2

Let A and B be non-empty sets. Let c be a constant. We must prove that if $c \notin A \cup B$, then $c \notin A \cap B$. You can draw a diagram if you wish; an example is shown in Fig. 6-4. The cross-hatched region represents $A \cap B$, and the shaded region represents $A \cup B$. Note the position of the point representing c.

We are told that $c \notin A \cup B$, that is, c is not an element of the union of sets A and B. From the definition of set union, it is not true that c is an element of set A or set B. DeMorgan's Law for disjunction, which was stated in Chapter 1, tells us that it is not true that c is an element of set A, and it is not true that c is an element of set B. That is, c is an element of neither set A nor set B. Indirectly, from the definition of set intersection, it follows that c is not an element of $A \cap B$. (The definition tells us that c can be an element of $A \cap B$ only if c is an element of both set A and set B, and that is clearly not the case.) Table 6-2 is an S/R version of this proof.

There's another way to solve this problem. What do you suppose it is? If you haven't guessed already, wait until you get to the quiz at the end of this chapter. Here are two hints. First, the statement $c \notin A \cup B$ can be rewritten as $\neg(c \in A \cup B)$, and the statement $c \notin A \cap B$ can be rewritten as $\neg(c \in A \cap B)$. Second, treat Solution 6-1 as a theorem, and apply one of the rules of propositional logic directly to it.

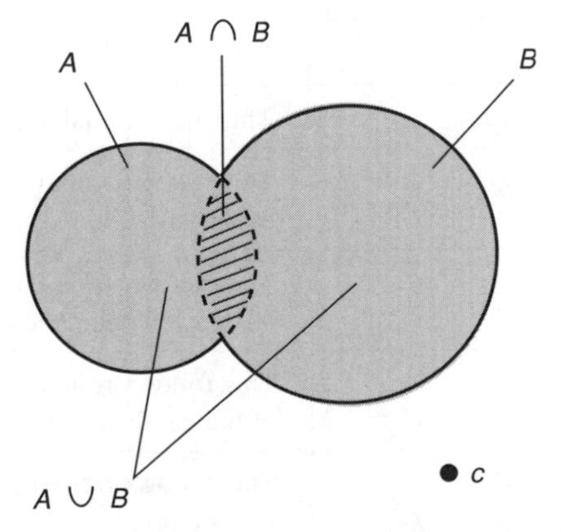

Fig. 6-4. Illustration for Problem 6-2.

Table 6-2. An S/R version of the proof demonstrated in Solution 6-2 and Fig. 6-4.

Statements	Reasons
Let A and B be non-empty sets.	We will use these in the proof.
Let c be a constant.	We will use this in the proof.
Assume $c \notin A \cup B$.	This is our initial assumption.
$\neg[(c \in A) \vee (c \in B)]$.	This follows from the definition of set union.
$\neg(c \in A)$ & $\neg(c \in B)$.	This follows from DeMorgan's Law for disjunction.
$(c \notin A)$ & $(c \notin B)$.	This is simply another way of stating the previous line.
$c \notin A \cap B$.	This follows from the definition of set intersection.

PROBLEM 6-3

Prove that if an element is in the union of two infinite sets, then it is not necessarily in their intersection.

SOLUTION 6-3

The key word in this problem is "necessarily." There are plenty of situations in which an element in the union of two infinite sets is also in their intersection. (Feel free to find some if you want.) But there are also situations in which an element in the union of two infinite sets is not in their intersection. Our mission is simply to find one example of a pair of infinite sets, call them S and T, and an element u such that $u \in S \cup T$, but $u \notin S \cap T$.

Let S be the set of even positive integers, and let T be the set of odd positive integers. That is:

$$S = \{2, 4, 6, 8, 10, \ldots\}$$
$$T = \{1, 3, 5, 7, 9, \ldots\}$$

In this case, the union of S and T is the set of all positive integers:

$$S \cup T = \{1, 2, 3, 4, 5, \ldots\}$$

These two sets are disjoint. Thus, by definition, their intersection is the null set:

$$S \cap T = \varnothing$$

A specific example of u can be chosen; any positive integer will do. Then $u \in S \cup T$, but $u \notin S \cap T$. But in fact, no matter what u might happen to be—positive integer, negative integer, rational number, irrational number, or anything else—we can be sure that it is not an element of $S \cap T$! (You'll get a chance to provide the reason for this in Quiz Question 4 at the end of this chapter.) Stated in logical symbols:

$$(\forall u)\ u \notin S \cap T$$

Having found an element u, a set S, and a set T such that that $u \in S \cup T$ and $u \notin S \cap T$, we can conclude the full statement that was to be proven:

$$(\exists u)(\exists S)(\exists T)\ [(u \in S \cup T)\ \&\ (u \notin S \cap T)]$$

In this usage, "but" is the logical equivalent of "and." Table 6-3 is an S/R version of this proof.

Table 6-3. An S/R version of the proof demonstrated in Solution 6-3.

Statements	**Reasons**
Let $S = \{2, 4, 6, 8, 10,\ldots\}$.	This is an example that sets the scene for the proof.
Let $T = \{1, 3, 5, 7, 9,\ldots\}$.	This is an example that sets the scene for the proof.
$S \cup T = \{1, 2, 3, 4, 5,\ldots\}$	This is apparent by examining sets S and T and denoting their elements in an increasing sequence.
Sets S and T are disjoint.	This follows from the fact that no even positive integer is odd, and the fact that no odd positive integer is even.
$S \cap T = \varnothing$.	This follows from the definition of disjoint sets.
$(\forall u)\ u \notin S \cap T$.	You'll get a chance to fill this in later.
$(\exists u)(\exists S)(\exists T)$ $[(u \in S \cup T)\ \&\ (u \notin S \cap T)]$.	We just found an example of an element u, a set S, and a set T for which $u \in S \cup T$ and $u \notin S \cap T$.

PROBLEM 6-4

Prove that the set of integers is a proper subset of the set of rational numbers.

SOLUTION 6-4

We must show two things. First, we have to prove that for every *p*, if *p* is an element of the set of integers, then *p* is an element of the set of rational numbers. Second, we must show that there exists some *q*, such that *q* is an element of the set of rational numbers but not an element of the set of integers. Symbolically, we can write these two propositions as follows:

$$\forall p\ [(p \in \boldsymbol{Z}) \Rightarrow (p \in \boldsymbol{Q})]$$
$$\exists q\ [(q \in \boldsymbol{Q})\ \&\ (q \notin \boldsymbol{Z})]$$

where ***Z*** is the set of integers and ***Q*** is the set of rational numbers.

The first proof is straightforward. Suppose that *p* is an element of set ***Z***. Invoking the division-by-1 axiom, we can claim that $p/1 = p$. By definition, 1 is an element of the set of integers. We know this because 1 is an element of the set $\{\ldots,-3, -2, -1, 0, 1, 2, 3,\ldots\}$, which is the set ***Z*** of integers. Thus *p* is equal to the quotient of two integers, because in the expression $p/1$, both the numerator and the denominator are integers. From the definition of rational number, it follows that *p* is an element of ***Q***, the set of rational numbers. Table 6-4A is an S/R version of this part of the proof.

Table 6-4A. An S/R version of the first part of the proof demonstrated in Solution 6-4.

Statements	Reasons
Let $p \in \boldsymbol{Z}$.	This is our initial assertion.
$p/1 = p$.	This follows from the division-by-1 axiom.
$1 \in \boldsymbol{Z}$.	This follows from the definition of the set of integers.
p is equal to the quotient of two integers with a nonzero denominator.	In the expression $p/1$, both the numerator and denominator are integers, and $1 \neq 0$.
$p \in \boldsymbol{Q}$.	This follows from the definition of rational number.

Table 6-4B. An S/R version of the second part of the proof demonstrated in Solution 6-4.

Statements	Reasons
Let $q = ½$.	This is our initial assertion.
$q \in \mathbf{Q}$.	We know that q is a rational number because ½ is equal to the quotient of two integers, namely 1 and 2.
$½ \notin \mathbf{Z}$.	We know this because $0 < ½ < 1$, and there exist no elements of **Z** between 0 and 1.
$q \notin \mathbf{Z}$.	This follows from the fact that $q = ½$, and ½ is not an integer.

In order to do the second part of the proof, we need to find a rational number that is not an integer. Suppose $q = ½$. By definition, q is a rational number, because ½ is equal to the quotient of two integers, namely 1 and 2. But ½ is not an element of the set **Z** of integers. Remember:

$$\mathbf{Z} = \{\ldots,-3, -2, -1, 0, 1, 2, 3,\ldots\}$$

When the elements of **Z** are listed this way, each element in the list is smaller than the one that immediately follows it, and there are no integers in between any two adjacent elements:

$$\ldots < -3 < -2 < -1 < 0 < 1 < 2 < 3 \ldots$$

The number $q = ½$ is not in this list, because $0 < ½ < 1$. Therefore, it is not true that q is an element of the set **Z**. Stated in set symbology, $q \notin \mathbf{Z}$. Table 6-4B is an S/R version of this part of the proof.

PROBLEM 6-5

Prove that the set of rational numbers is a proper subset of the set of real numbers.

SOLUTION 6-5

This proof, like the one in the preceding problem, consists of two parts. First, we have to prove that for every r, if r is an element of the set of rational numbers, then r is an element of the set of real numbers. Second, we must show that there exists some s, such that s is an element of the set of real numbers but not an element of the set of rational numbers. Symbolically, we can write these two propositions as follows:

$$\forall r\ [(r \in \boldsymbol{Q}) \Rightarrow (r \in \boldsymbol{R})]$$
$$\exists s\ [(s \in \boldsymbol{R})\ \&\ (s \notin \boldsymbol{Q})]$$

where $\boldsymbol{Q}$ is the set of rational numbers and $\boldsymbol{R}$ is the set of real numbers.

The first part of this proof is straightforward. Let $r \in \boldsymbol{Q}$. From the definition of logical disjunction, r is an element of set $\boldsymbol{Q}$ or r is an element of set $\boldsymbol{X}$, where $\boldsymbol{X}$ can represent any set whatsoever. It follows that r is an element of the union of sets $\boldsymbol{Q}$ and $\boldsymbol{X}$. (You'll get a chance to provide the reason for this in Quiz Question 5 at the end of this chapter.) Consider set $\boldsymbol{X}$ to be the set $\boldsymbol{S}$ of irrational numbers. We can do this because $\boldsymbol{X}$ can be any set we want. Then r is an element of the union of sets $\boldsymbol{Q}$ and $\boldsymbol{S}$. According to the definition of the set of real numbers, the union of sets $\boldsymbol{Q}$ and $\boldsymbol{S}$ is the set $\boldsymbol{R}$. Therefore, r is an element of set $\boldsymbol{R}$. Table 6-5A is an S/R version of this part of the proof.

Table 6-5A. An S/R version of the first part of the proof demonstrated in Solution 6-5.

Statements	Reasons
Let $r \in \boldsymbol{Q}$.	This is our initial assertion.
$(r \in \boldsymbol{Q}) \vee (r \in \boldsymbol{X})$.	This follows from the definition of logical disjunction, where $\boldsymbol{X}$ can be any set whatsoever.
$r \in \boldsymbol{Q} \cup \boldsymbol{X}$.	You'll get a chance to supply the reason for this later.
Consider $\boldsymbol{X} = \boldsymbol{S}$.	We can do this because $\boldsymbol{X}$ can be any set we choose.
$r \in \boldsymbol{Q} \cup \boldsymbol{S}$.	This follows from the previous two steps.
$\boldsymbol{Q} \cup \boldsymbol{S} = \boldsymbol{R}$.	This follows from the definition of the set of real numbers.
$r \in \boldsymbol{R}$.	This follows from the previous two steps.

In order to do the second part of the proof, we must find a real number that is not a rational number. Let s be an irrational number; that is, let s be an element of set $\mathbf{S}$. From the definition of logical disjunction, s is an element of set $\mathbf{S}$ or s is an element of set $\mathbf{Y}$, where $\mathbf{Y}$ can represent any set whatsoever. It follows that s is an element of the union of sets $\mathbf{S}$ and $\mathbf{Y}$. (You'll get a chance to provide the reason for this in the chapter-ending quiz.) Consider set $\mathbf{Y}$ to be the set $\mathbf{Q}$ of rational numbers. We can do this because $\mathbf{Y}$ can be any set we want. Then s is an element of the union of sets $\mathbf{S}$ and $\mathbf{Q}$. According to the definition of the set of real numbers, the union of sets $\mathbf{S}$ and $\mathbf{Q}$ is the set $\mathbf{R}$. Therefore, s is an element of set $\mathbf{R}$. We know that s is not an element of the set $\mathbf{Q}$ of rational numbers. This follows from the definitions of irrational and rational numbers; s is not a ratio of two integers, but in order to be in set $\mathbf{Q}$, it would have to be. Table 6-5B is an S/R version of this part of the proof.

Table 6-5B. An S/R version of the second part of the proof demonstrated in Solution 6-5.

Statements	Reasons
Let $s \in \mathbf{S}$.	This is our initial assertion.
$(s \in \mathbf{S}) \vee (s \in \mathbf{Y})$.	This follows from the definition of logical disjunction, where $\mathbf{Y}$ can be any set whatsoever.
$s \in \mathbf{S} \cup \mathbf{Y}$.	You'll get a chance to supply the reason for this later.
Consider $\mathbf{Y} = \mathbf{Q}$.	We can do this because $\mathbf{Y}$ can be any set we choose.
$s \in \mathbf{S} \cup \mathbf{Q}$.	This follows from the previous two steps.
$\mathbf{S} \cup \mathbf{Q} = \mathbf{R}$.	This follows from the definition of the set of real numbers.
$s \in \mathbf{R}$.	This follows from the previous two steps.
$s \notin \mathbf{Q}$.	This follows from the definitions of irrational and rational numbers.
$(s \in \mathbf{R})$ & $(s \notin \mathbf{Q})$.	This follows directly from the previous two statements.

PROBLEM 6-6

Prove that the sum of two rational numbers is a rational number.

SOLUTION 6-6

Let r be a rational number such that $r = a/b$, where a and b are integers and $b \neq 0$. Let s be a rational number such that $s = c/d$, where c and d are integers and $d \neq 0$. From the definition of a rational number, we know such integers exist for all rational numbers r and s. The sum of r and s can be written as follows:

$$r + s = a/b + c/d$$

From the sum-of-fractions axiom, it follows that:

$$r + s = (ad + bc)/bd$$

Let $ad = e$, let $bc = f$, and let $bd = g$, renaming these products for simplicity. From the product-of-integers axiom, we know that e, f, and g are integers. Substituting in the above equation:

$$r + s = (e + f)/g$$

Let $e + f = h$. (We rename this sum for simplicity.) We know that h is an integer. (You'll get a chance to provide the reason for this in Quiz Question 6 at the end of this chapter.) Substituting in the above equation:

$$r + s = h/g$$

where h and g are both integers. By applying DeMorgan's Law for disjunction to the second part of the product-of-integers axiom, we can conclude that $g \neq 0$, because $g = bd$, $b \neq 0$, and $d \neq 0$. By definition, then, h/g is a rational number, because it is the quotient of two integers with a nonzero denominator. By substitution, $r + s$ is the quotient of two integers with a nonzero denominator. Therefore, according to the definition of a rational number, $r + s$ is an element of the set $\boldsymbol{Q}$ of rational numbers. Table 6-6 is an S/R version of this proof.

PROBLEM 6-7

Prove that the difference between two rational numbers is a rational number.

Table 6-6. An S/R version of the proof demonstrated in Solution 6-6.

Statements	**Reasons**
Let $r \in \boldsymbol{Q}$ and $s \in \boldsymbol{Q}$.	We will use these in the proof.
Let $a \in \mathbf{Z}$, $b \in \mathbf{Z}$, $c \in \mathbf{Z}$, $d \in \mathbf{Z}$, $b \neq 0$, and $d \neq 0$.	We will use these in the proof.
Let $r = a/b$.	This is defined because $b \neq 0$.
Let $s = c/d$.	This is defined because $d \neq 0$.
$r + s = a/b + c/d$.	This is a mere matter of substitution.
$r + s = (ab + bc)/bd$.	This follows from the sum-of-fractions axiom.
Let $ad = e$, let $bc = f$, and let $bd = g$.	Rename these products for simplicity.
$e \in \mathbf{Z}, f \in \mathbf{Z}$, and $g \in \mathbf{Z}$.	This follows from the product-of-integers axiom.
$r + s = (e + f)/g$.	This is a mere matter of substitution.
Let $e + f = h$.	Rename this sum for simplicity.
$h \in \mathbf{Z}$.	You'll get a chance to fill this in later.
$r + s = h/g$.	This is a mere matter of substitution.
$g \neq 0$.	This is because $g = bd$ with $b \neq 0$ and $d \neq 0$, so the product-of-integers axiom, with the help of DeMorgan's Law for disjunction, ensures that $g \neq 0$.
h/g is the quotient of two integers with nonzero denominator.	We already know that $g \in \mathbf{Z}$, $h \in \mathbf{Z}$, and $g \neq 0$.
$r + s$ is the quotient of two integers with nonzero denominator.	This is a mere matter of substitution.
$(r + s) \in \boldsymbol{Q}$.	This follows from the definition of a rational number.

SOLUTION 6-7

Let r be a rational number such that $r = a/b$, where a and b are integers and $b \neq 0$. Let s be a rational number such that $s = c/d$, where c and d are integers and $d \neq 0$. From the definition of a rational number, we know such integers exist for all rational numbers r and s. By substituting a/b for r, substituting c/d for s, invoking the difference-between-fractions axiom, and then substituting r in place of a/b, the difference between r and s can be rewritten as follows:

$$\begin{aligned} r - s &= a/b - c/d \\ &= a/b + (-c/d) \\ &= r + (-c/d) \end{aligned}$$

We know that $-c = -1 \times c$. (You'll get a chance to provide the reason in Quiz Question 7 at the end of this chapter.) Because -1 and c are both integers, it follows from the product-of-integers axiom that $-c$ is an integer. Therefore, $-c/d$ is the quotient of two integers with a nonzero denominator. By definition, $-c/d$ is a rational number. We are given the fact that r is a rational number. From the result of Solution 6-6, we know that the sum of two rational numbers is always rational, so $r + (-c/d)$ is rational. We have already determined that $r - s = r + (-c/d)$. It follows that $r - s$ is a rational number. Table 6-7 is an S/R version of this proof.

PROBLEM 6-8

Prove that the product of two rational numbers is a rational number.

SOLUTION 6-8

Let r be a rational number such that $r = a/b$, where a and b are integers and $b \neq 0$. Let s be a rational number such that $s = c/d$, where c and d are integers and $d \neq 0$. From the definition of a rational number, we know such integers exist for all rational numbers r and s. By substituting a/b for r, substituting c/d for s, and then invoking the product-of-fractions axiom, the product rs can be rewritten as follows:

$$\begin{aligned} rs &= (a/b)(c/d) \\ &= (ac)/(bd) \end{aligned}$$

According to the product-of-integers axiom, we know that ac is an integer, and that bd is an integer. By applying DeMorgan's Law for disjunction to the second part of the product-of-integers axiom, we can conclude that $bd \neq 0$, because $b \neq 0$ and $d \neq 0$. This means $(ac)/(bd)$ is the quotient of two integers with a nonzero denominator. It follows

Table 6-7. An S/R version of the proof demonstrated in Solution 6-7.

Statements	**Reasons**
Let $r \in \boldsymbol{Q}$ and $s \in \boldsymbol{Q}$.	We will use these in the proof.
Let $a \in \mathbf{Z}$, $b \in \mathbf{Z}$, $c \in \mathbf{Z}$, $d \in \mathbf{Z}$, $b \neq 0$, and $d \neq 0$.	We will use these in the proof.
Let $r = a/b$.	This is defined because $b \neq 0$.
Let $s = c/d$.	This is defined because $d \neq 0$.
$r - s = a/b - c/d$.	This is the result of substituting a/b for r, and c/d for s.
$a/b - c/d = a/b + (-c/d)$.	This follows from the difference-between-fractions axiom.
$a/b + (-c/d) = r + (-c/d)$.	This is the result of substituting r for a/b in the previous statement.
$r - s = r + (-c/d)$.	This follows from the previous three steps.
$-c = -1 \times c$.	You'll get a chance to fill this in later.
$-1 \in \mathbf{Z}$ and $c \in \mathbf{Z}$.	For -1, the fact follows from the definition of integer. For c, we are told this at the outset.
$-c \in \mathbf{Z}$.	This follows from the previous two steps and the product-of-integers axiom.
$-c/d$ is the quotient of two integers, and the denominator is nonzero.	This follows from the previous step and the outset, where we are told that $d \in \mathbf{Z}$ and also that $d \neq 0$.
$-c/d \in \boldsymbol{Q}$.	This follows from the definition of a rational number.
$r \in \boldsymbol{Q}$.	We are told this at the outset.
$r + (-c/d) \in \boldsymbol{Q}$.	This follows from the previous two steps and from Solution 6-6.
$r - s = r + (-c/d)$.	This was previously determined.
$r - s \in \boldsymbol{Q}$.	This follows from the previous two steps.

Table 6-8. An S/R version of the proof demonstrated in Solution 6-8.

Statements	Reasons
Let $r \in \boldsymbol{Q}$ and $s \in \boldsymbol{Q}$.	We will use these in the proof.
Let $a \in \mathbf{Z}$, $b \in \mathbf{Z}$, $c \in \mathbf{Z}$, $d \in \mathbf{Z}$, $b \neq 0$, and $d \neq 0$.	We will use these in the proof.
Let $r = a/b$.	This is defined because $b \neq 0$.
Let $s = c/d$.	This is defined because $d \neq 0$.
$rs = (a/b)(c/d)$.	This is the result of substituting a/b for r, and c/d for s.
$(a/b)(c/d) = (ac)/(bd)$.	This follows from the product-of-fractions axiom.
$rs = (ac)/(bd)$.	This follows from the previous two steps.
$ac \in \boldsymbol{Q}$ and $bd \in \boldsymbol{Q}$.	This follows from the product-of-integers axiom.
$bd \neq 0$.	This is because $b \neq 0$ and $d \neq 0$, so the product-of-integers axiom, with the help of DeMorgan's Law for disjunction, ensures that $bd \neq 0$.
$(ac)/(bd)$ is the quotient of two integers with a nonzero denominator.	This follows directly from the previous step.
$(ac)/(bd) \in \boldsymbol{Q}$.	You'll get a chance to fill this in later.
$rs \in \boldsymbol{Q}$.	This is the result of substituting rs for the quantity $(ac)/(bd)$.

that $(ac)/(bd)$ is a rational number. (You'll get a chance to provide the reason for this in Quiz Question 8 at the end of this chapter.) We have already determined that $rs = (ac)/(bd)$. By substitution, it follows that rs is a rational number. Table 6-8 is an S/R version of this proof.

PROBLEM 6-9

Prove that the quotient of two rational numbers is a rational number, if the denominator is not equal to 0.

SOLUTION 6-9

Let r be a rational number such that $r = a/b$, where a and b are integers and $b \neq 0$. Let s be a rational number such that $s = c/d$, where c and d are integers with $c \neq 0$ and $d \neq 0$. From the definition of a rational number, we know such integers a and b exist for all rational numbers r, and that such integers c and d exist for all nonzero rational numbers s. By substituting a/b for r, substituting c/d for s, and then invoking the quotient-of-fractions axiom, the quotient r/s can be rewritten as follows:

$$\begin{aligned} r/s &= (a/b)/(c/d) \\ &= (ad)/(bc) \end{aligned}$$

According to the product-of-integers axiom, we know that ad is an integer, and also that bc is an integer. By applying DeMorgan's Law for disjunction to the second part of the product-of-integers axiom, we can conclude that $bc \neq 0$, because $b \neq 0$ and $c \neq 0$. This means $(ad)/(bc)$ is the quotient of two integers with a nonzero denominator. It follows that $(ad)/(bc)$ is a rational number. (You'll get a chance to provide the reason for this in Quiz Question 8 at the end of this chapter.) We have already determined that $r/s = (ad)/(bc)$. By substitution, we can conclude that r/s is a rational number. Table 6-9 is an S/R version of this proof.

PROBLEM 6-10

Prove that for all real numbers r, s, t, and u:

$$(r + s)(t + u) = rt + ru + st + su$$

SOLUTION 6-10

The proof of this proposition is really nothing more than an algebraic derivation in which we must justify each step. In order to get started, consider the quantity $(r + s)$ as a single value. Then, according to the distributive axiom, the above equation can be rewritten as follows:

$$(r + s)(t + u) = (r + s)t + (r + s)u$$

We can rearrange it further to get:

$$(r + s)t + (r + s)u = t(r + s) + u(r + s)$$

(You'll get a chance to provide the reason for this in Quiz Question 9.) Applying the distributive axiom again, we can derive the following:

$$t(r + s) + u(r + s) = tr + ts + ur + us$$

Table 6-9. An S/R version of the proof demonstrated in Solution 6-9.

Statements	Reasons
Let $r \in \mathbf{Q}$ and $s \in \mathbf{Q}$.	We will use these in the proof.
Let $a \in \mathbf{Z}$, $b \in \mathbf{Z}$, $c \in \mathbf{Z}$, $d \in \mathbf{Z}$, $b \neq 0$, $c \neq 0$, and $d \neq 0$.	We will use these in the proof.
Let $r = a/b$.	This is defined because $b \neq 0$.
Let $s = c/d$.	This is defined because $d \neq 0$.
$r/s = (a/b)/(c/d)$.	This is the result of substituting a/b for r, and c/d for s.
$(a/b)/(c/d) = (ad)/(bc)$.	This follows from the quotient-of-fractions axiom.
$r/s = (ad)/(bc)$.	This follows from the previous two steps.
$ad \in \mathbf{Q}$ and $bc \in \mathbf{Q}$.	This follows from the product-of-integers axiom.
$bc \neq 0$.	This is because $b \neq 0$ and $c \neq 0$, so the product-of-integers axiom, with the help of DeMorgan's Law for disjunction, ensures that $bc \neq 0$.
$(ad)/(bc)$ is the quotient of two integers with a nonzero denominator.	This follows directly from the previous step.
$(ad)/(bc) \in \mathbf{Q}$.	You'll get a chance to fill this in later.
$r/s \in \mathbf{Q}$.	This is the result of substituting r/s for the quantity $(ad)/(bc)$.

By invoking the commutative axiom for multiplication four times (once for each addend), the equation can be further modified to obtain:

$$tr + ts + ur + us = rt + st + ru + su$$

Finally, according to the commutative axiom for addition, we derive the desired expression:

$$rt + st + ru + su = rt + ru + st + su$$

Table 6-10. An S/R version of the proof demonstrated in Solution 6-10.

Statements	**Reasons**
Consider the quantity $(r + s)$ as a single value.	We need to do this to get started!
$(r + s)(t + u)$ $= (r + s)t + (r + s)u$	This follows from the distributive axiom.
$(r + s)\text{t} + (r + s)u$ $= t(r + s) + u(r + s)$	You'll get a chance to fill this in later.
$t(r + s) + u(r + s)$ $= tr + ts + ur + us$	This follows from the distributive axiom.
$tr + ts + ur + us$ $= rt + st + ru + su$	This follows from the commutative axiom for multiplication.
$rt + st + ru + su$ $= rt + ru + st + su$	This follows from the commutative axiom for addition.
$(r + s)(t + u)$ $= rt + ru + st + su$	This follows from repeated application of a component of the equality axiom. You will get a chance to identify the component later.

All of the previous equations allow us to claim what was to be proved, on the basis of repeated application of a component (you'll have a chance to identify the component in Quiz Question 10) of the equality axiom:

$$(r + s)(t + u) = rt + ru + st + su$$

Table 6-10 is an S/R version of this proof.

Quiz

This is an "open book" quiz. You may refer to the text in this chapter. A good score is 8 correct. Answers are in the back of the book.

1. What is the cardinality of the empty set?
   (a) 0
   (b) 1
   (c) Infinity
   (d) It is not defined.

2. An equivalence relation is
   (a) symmetric.
   (b) distributive.
   (c) denumerable.
   (d) All of the above

3. Look back at Problems 6-1 and 6-2, and their solutions. Which rule of propositional logic can be applied directly to Solution 6-1 (treating Solution 6-1 as a theorem), in order to solve Problem 6-2 by the second method mentioned? Look back at Chapter 1 if you must.
   (a) DeMorgan's law for implication.
   (b) The law of implication reversal.
   (c) The commutative law for disjunction.
   (d) The commutative law for conjunction.

4. At the end of Solution 6-3 and in the next-to-last line of Table 6-3, you are told that you'll get a chance to provide the reason for the statement. Now is the time! What is the reason?
   (a) The element u has not been specified, so we cannot say it is in the intersection of two specific sets.
   (b) We are making a generalization about the element u, so we cannot say anything specific about it.
   (c) No matter what u happens to be, it can't be in a set that has no elements.
   (d) Sets S and T are not the same, so obviously element u cannot be in both of them.

5. Refer to Solution 6-5 and Tables 6-5A and 6-5B. In the first part of the proof and in Table 6-5A, one of the steps makes the claim that $r \in \boldsymbol{Q} \cup \boldsymbol{X}$. In the second part of the proof and in Table 6-5B, one of the steps says that $s \in \boldsymbol{S} \cup \boldsymbol{Y}$. The reason in both cases is the same, but it's left out, and you're told you'll get a chance to identify it. Your chance has come! What is the reason for this statement in both parts of the proof?
   (a) It follows from the definition of logical disjunction.
   (b) It follows from the definition of logical implication.
   (c) It follows from the definition of set intersection.
   (d) It follows from the definition of set union.

6. Refer to Solution 6-6 and Table 6-6. In this proof, one of the statements says that h is an integer (that is, $h \in \boldsymbol{Z}$). What allows us to make this claim?

(a) The product-of-integers axiom.
(b) The product-of-fractions axiom.
(c) The sum-of-integers axiom.
(d) The sum-of-fractions axiom.

7. Refer to Solution 6-7 and Table 6-7. One of the statements says that $-c = -1 \times c$. What allows us to make this claim?

    (a) The additive-inverse axiom.
    (b) The sum-of-integers axiom.
    (c) The product-of-integers axiom.
    (d) The definition of a rational number.

8. Refer to Solution 6-8 and Table 6-8. One of the statements says that $(ac)/(bd) \in \boldsymbol{Q}$. A similar situation occurs in Solution 6-9 and Table 6-9, where one of the statements says that $(ad)/(bc) \in \boldsymbol{Q}$. What allows us to make these claims?

    (a) The additive-inverse axiom.
    (b) The sum-of-integers axiom.
    (c) The product-of-integers axiom.
    (d) The definition of a rational number.

9. Refer to Solution 6-10 and Table 6-10. One of the statements says the following:

$$(r + s)t + (r + s)u = t(r + s) + u(r + s)$$

    After this statement, you're told you'll get a chance to provide the reason. Your chance has come! What allows us to make this claim?

    (a) The commutative axiom for addition, applied twice.
    (b) The commutative axiom for multiplication, applied twice.
    (c) The equality axiom.
    (d) The distributive axiom.

10. Refer again to Solution 6-10 and Table 6-10. In the last step, it says that a certain component of the equality axiom is used repeatedly to come to the conclusion shown. You're told you'll get a chance to identify that component. Your chance has come! Which component is it?

    (a) The reflexive property.
    (b) The symmetric property.
    (c) The transitive property.
    (d) The equivalence property.

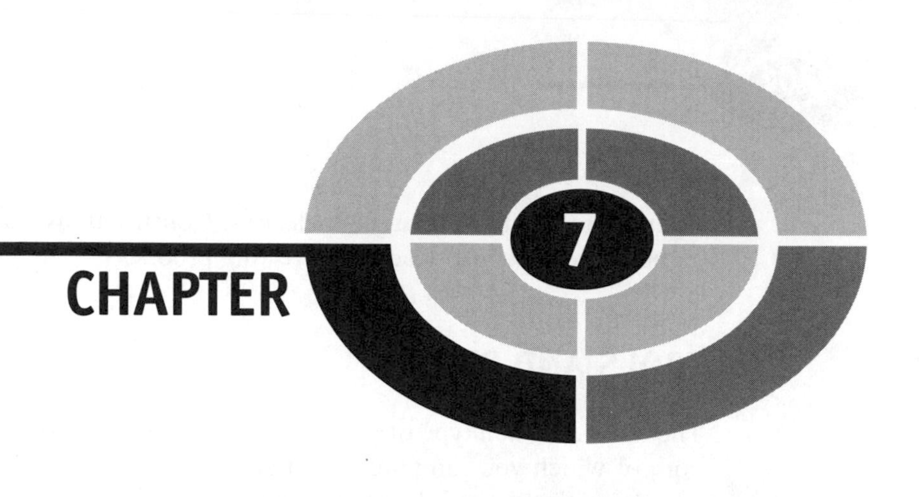

A Few Historic Tidbits

If you've carefully followed this book up to now, you should have a good idea of how pure mathematics "works." Let's look at a few ideas and proofs that are significant in the history of mathematics. This chapter is only a tiny sampler. Some great theorems, such as Gödel's theorems about logical systems, Cantor's proofs involving the existence of different magnitudes of infinity, and Fermat's Last Theorem are not included because their proofs are beyond the scope of this book.

You "Build" It

In geometry, a *construction* is a drawing made with simple instruments. Constructions make you think about the properties of geometric objects, independent of numeric lengths and angle measures. Constructions can be great intellectual games. They can also lead to interesting proofs.

TOOLS AND RULES

The most common type of geometric construction is done with two instruments, both of which you can purchase at any office supply store. One instrument lets you draw circles, and the other lets you draw straight line segments. Once you have these, you can use them only according to certain "rules of the game."

DRAFTING COMPASS

The *drafting compass* is an uncalibrated device for drawing circles of various sizes. It has two straight shafts joined at one end with a hinge. One shaft ends in a sharp point that does not mark anything, but that can be stuck into a piece of paper as an anchor. The other shaft has brackets in which a pen or pencil is mounted. To draw a circle, press the sharp point down on a piece of paper (with some cardboard underneath to protect the table or desk top), open the hinge to get the desired radius, and draw the circle by rotating the whole assembly at least once around. You can draw arcs by rotating the compass partway around.

For geometric constructions, the ideal compass does not have an angle measurement ("degree") scale at its hinge. If it has a scale that indicates angle measures or otherwise quantifies the extent to which it is opened, you must ignore that scale.

STRAIGHT EDGE

A *straight edge* is any object that helps you to draw line segments by placing a pen or pencil against the object and running it alongside. A conventional ruler will work for this purpose, but is not the best tool to use because it has a calibrated scale. A better tool is a strip of wood, cardboard, or metal having a convenient length. A drafting triangle can also be used if it is uncalibrated, but you must ignore the angles at the apexes. Some drafting triangles have two 45° angles and one 90° angle; others have one 30° angle, one 60° angle, and one 90° angle. You aren't allowed to take advantage of these standard angle measures when performing geometric constructions.

WHAT'S ALLOWED

With a compass, you can draw circles or arcs having any radius you want. The center point can be randomly chosen, or you can place the sharp tip of the compass down on a predetermined, existing point and make it the center point of the circle or arc.

You can set a compass to replicate the distance between any two defined points, by setting the non-marking tip down on one point and the marking tip down on the other point, and then holding the compass setting constant.

With the straight edge, you can connect any two specific points with a line segment. You can extend this line segment beyond the points, denoting the line passing through those two points. You can also draw a line in any direction, either "randomly" (by simply setting the straight edge down and passing the pen or pencil along it), or through a specific point. You can draw a ray through any two specific points, as long as one of the points is the end point of the ray. You can draw a ray emanating in any direction from a single point, where that point constitutes the end point of the ray.

WHAT'S FORBIDDEN

Whatever circle or line segment you draw, you are not allowed to measure the radius or the length against a calibrated scale. You may not measure angles using a calibrated device. You may not make reference marks on the compass or the straight edge. (Marking on a straight edge is "cheating," even though replicating the distance between two points with a compass is all right.) You may not extend an existing line segment using the straight edge, except when you draw a ray or a line based on the same two points you use to define that line segment.

Here is a subtle but important restriction: You may not make use of the implied result of an infinite number of repetitions of a single operation or sequence of operations. For example, you are not allowed to mentally perform a maneuver over and over *ad infinitum* to geometrically approach a desired result, and then claim that result as a valid construction. The entire operation must be completed in a finite number of steps.

DEFINING POINTS

To define an arbitrary point, all you need to do is draw a little dot on the paper. Alternatively, you can set the non-marking point of the compass down on the paper, in preparation for drawing an arc or circle centered at an arbitrary point. Points can also be defined where two line segments intersect, where an arc or circle intersects a line segment, or where an arc or circle intersects another arc or circle.

DRAWING LINE SEGMENTS

When you want to draw a line segment through two defined points (call them P and Q), place the tip of the pencil on one of the points and place the straight edge down against the point of the pencil. Rotate the straight edge until it lines up with the other point while still firmly resting against the tip of the pencil, and then run the pencil back and forth along the edge, so the mark connects both points. Be sure the pencil makes its mark only between the points, and not past them on either side (Fig. 7-1). Of course, you have to use a straight edge that's at least as long as the distance between the two points!

DENOTING RAYS

In order to denote a ray, first determine or choose the end point P of the ray. Then place the tip of the pencil at the end point, and place the straight edge against the tip of the pencil. Orient the straight edge so it runs through the other point Q that defines the ray. Move the tip of the pencil away from the point in the direction of the ray, as far as you want without running off the end of the

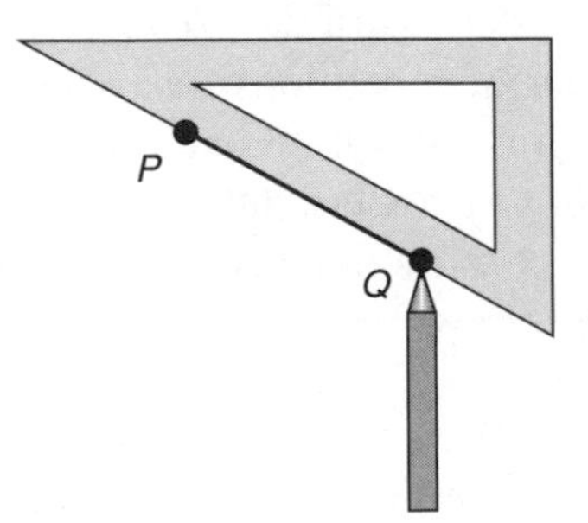

Fig. 7-1. Construction of a line segment connecting two predetermined points.

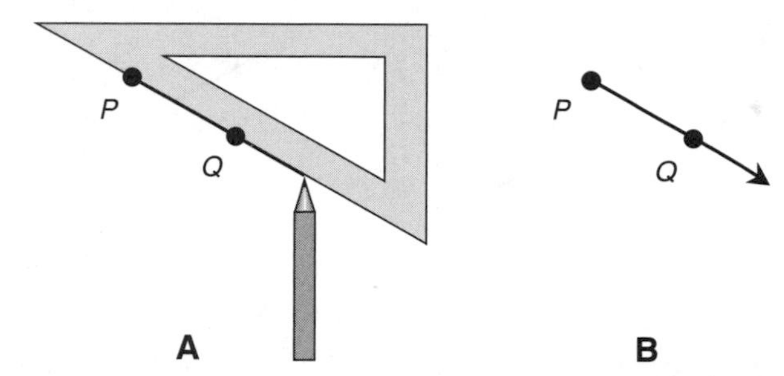

Fig. 7-2. Construction of a ray. First draw a solid straight mark starting at a point P and passing through a point Q (as shown at A); then put an arrow at the end of the straight mark opposite the point P (as shown at B).

straight edge (Fig. 7-2A). Finally, draw an arrow at the end of the straight mark you have made, opposite the starting point (Fig. 7-2B). The arrow indicates that the ray extends infinitely in that direction.

DENOTING LINES

A line can be drawn at random through no point in particular (as shown at Figs. 7-3A and 7-3B), at random through a single defined point (as shown at C and D), or specifically through two defined points (as shown at E and F). In order to draw a line through two points, follow the same procedure as you would to draw a line segment, but extend the mark past the two points in both directions. Then place arrows at both ends of the mark.

DRAWING CIRCLES

To draw a circle around a random point, place the non-marking tip of the compass down on the paper, set the compass to the desired radius, and rotate the instrument through a full circle (Fig. 7-4A). If the center point is predetermined (marked by a dot), place the non-marking tip down on the dot and rotate the instrument through a full circle.

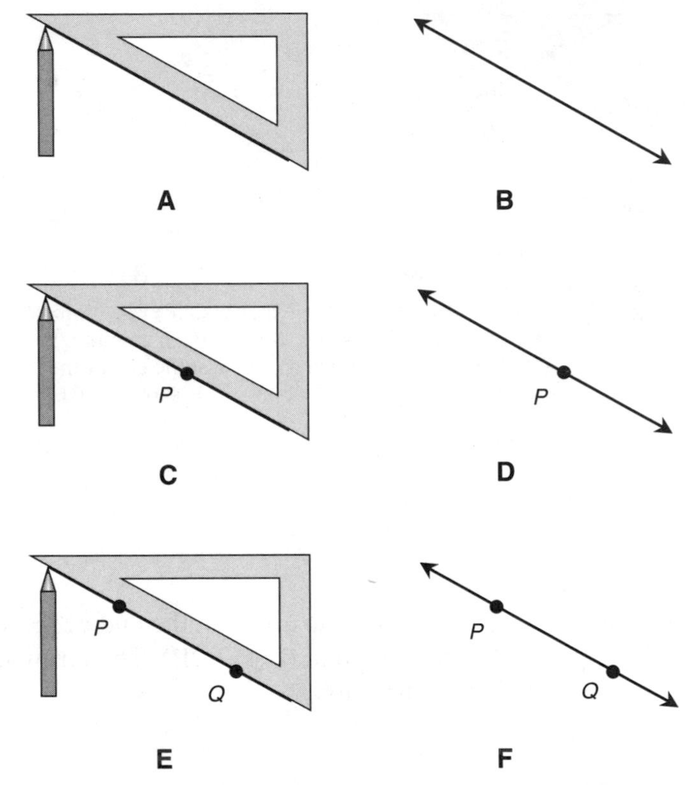

Fig. 7-3. At A and B, construction of an arbitrary line. At C and D, construction of a line through a single predetermined point. At E and F, construction of a line through two predetermined points.

DRAWING ARCS

To draw an arc centered at a random point, place the non-marking tip of the compass down on the paper, set the compass to the desired radius, and rotate the instrument through the desired arc. If the center point is predetermined (marked by a dot), place the non-marking tip down on the dot and rotate the instrument through the desired arc (Fig. 7-4B).

PROBLEM 7-1

Define a point by drawing a dot. Then, with the compass, draw a small circle centered on the dot. Now construct a second circle, concentric with the first one, but having twice the radius.

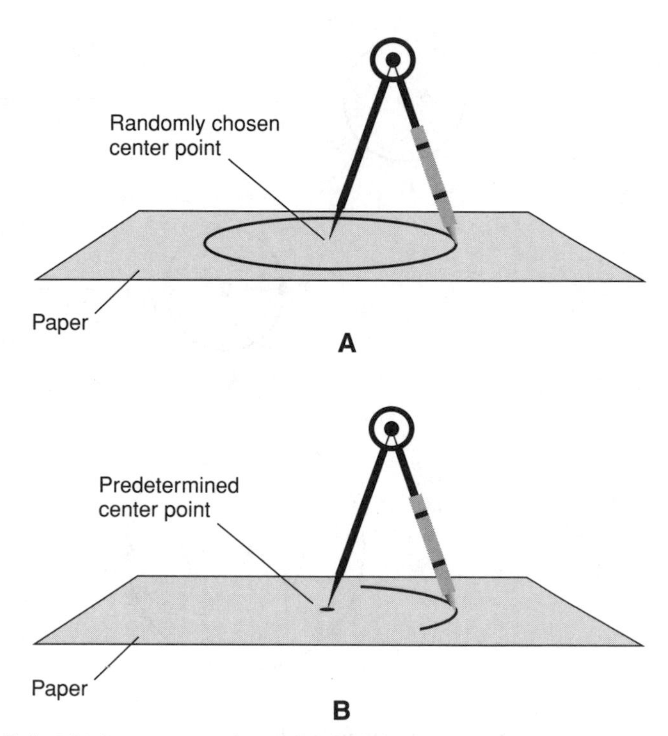

Fig. 7-4. At A, a compass is used to draw a circle around a randomly chosen center point. At B, a compass is used to draw an arc centered at a predetermined point.

SOLUTION 7-1

Fig. 7-5 illustrates the procedure. In drawing A, the circle is constructed with the compass, centered at the initial point (called point P). In draw ing B, a ray L is drawn using the straight edge, with one end at point P and passing through the circle at a point Q. In drawing C, a circle is constructed, centered at point Q and leaving the compass set for the same radius as it was when the original circle was drawn. This new circle intersects L at point P (the center of the original circle) and also at a new point R. Next, the non-marking tip of the compass is placed back at point P, and the compass is opened up so the pencil tip falls exactly on point R. Finally, as shown in drawing D, a new circle is drawn with its center at point P, with a radius equal to the length of line segment PR.

PROBLEM 7-2

Draw three points on a piece of paper, placed so they do not all lie along the same line. Label the points P, Q, and R. Construct ΔPQR

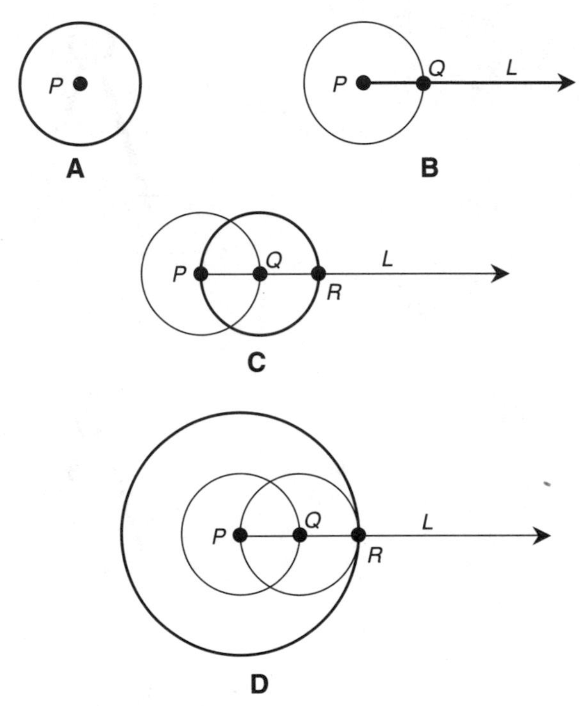

Fig. 7-5. Illustration for Problem 7-1.

connecting these three points. Draw a circle whose radius is equal to the length of side PQ, but that is centered at point R.

SOLUTION 7-2

The process is shown in Fig. 7-6. In drawing A, the three points are put down and labeled. In drawing B, the points are connected to form ΔPQR. Drawing C shows how the non-marking tip of the compass is placed at point Q, and the tip of the pencil is placed on point P. (You don't have to draw the arc, but it is included in this illustration for emphasis.) With the compass thus set so it defines the length of line segment PQ, the non-marking tip of the compass is placed on point R. Finally, as shown in drawing D, the circle is constructed.

PROBLEM 7-3

Can the non-marking tip of the compass be placed at point P, and the pencil tip placed to draw an arc through point Q, in order to define the length of line segment PQ in Fig. 7-6?

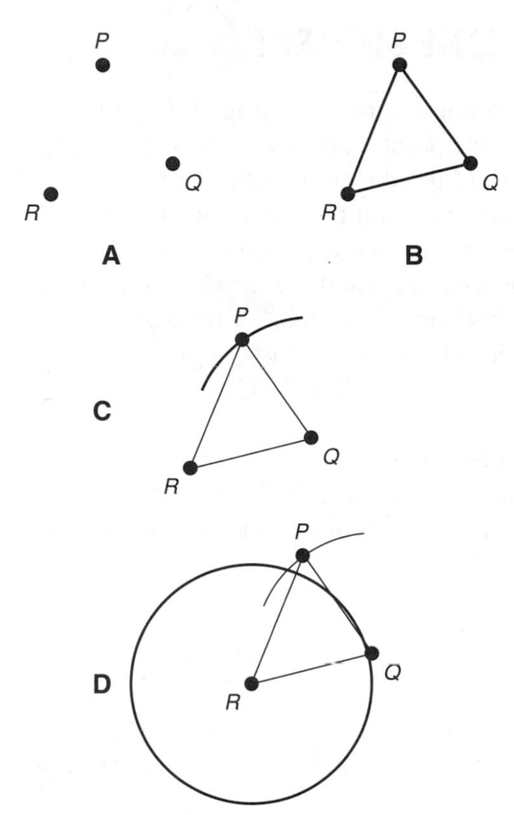

Fig. 7-6. Illustration for Problems 7-2 and 7-3. Note that point Q, while close to the circle, does not actually lie on the circle.

SOLUTION 7-3

Yes. This will work just as well.

BISECTION

A line segment is said to be *bisected* if and only if it is divided by a specific point into two shorter line segments of equal length. The bisecting point can exist all by itself, or it can represent the intersection of the original line segment with another line segment, a ray, or a line. An angle is said to be bisected if and only if it is divided into two smaller angles of equal measure, by a ray whose end point coincides with the vertex of the bisected angle.

BISECTING A LINE SEGMENT

Suppose you have a line segment *PQ* (Fig. 7-7A) and you want to find the point at its center, that is, the point that bisects line segment *PQ*. First, construct an arc centered at point *P*. Make the arc roughly half-circular, and set the compass to span somewhat more than half the length of *PQ*. Then, without altering the setting of the compass, draw an arc centered at point *Q*, such that its radius is the same as that of the first arc you drew (as shown in Fig. 7-7B). Name the points at which the two arcs intersect *R* and *S*. Construct a line passing through both *R* and *S*. Line *RS* intersects the original line segment *PQ* at a point *T*, which bisects line segment *PQ* (as shown in Fig. 7-7C).

PROBLEM 7-4

Prove that the above described method of bisecting a line segment is valid according to the rules of construction and the principles of geometry.

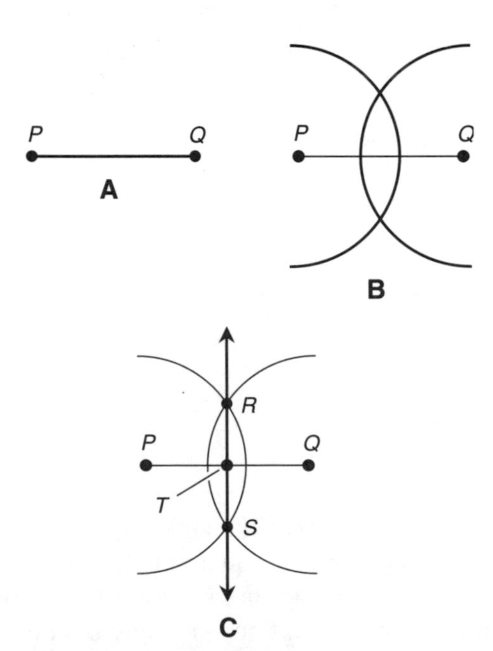

Fig. 7-7. Bisection of a line segment.

SOLUTION 7-4

You may wish to expand and add some more detail to Fig. 7-7C, obtaining a drawing that looks like Fig. 7-8. The proof can be split into two parts.

First, consider ΔSRP and ΔRSQ. We know that the lengths of the four line segments *RP*, *PS*, *SQ*, and *QR* are all equal, because they all represent the radius of the arcs, both of which were created with the compass set for the same span. This means that ΔSRP and ΔRSQ are isosceles triangles, and two pairs of corresponding sides have equal lengths. They also have line segment *RS* in common, so all three pairs of corresponding sides, as we proceed in the same direction around either triangle, are equal in length. Invoking the SSS axiom, we can conclude that ΔSRP is directly congruent to ΔRSQ. Now we know that the measures of the corresponding pairs of angles, as we proceed around both triangles in the same direction, are equal. (You'll get a chance to provide the reason for this in Quiz Question 1 at the end of this chapter.) This, in addition to the fact that ΔSRP and ΔRSQ are isosceles triangles, means that the measures of $\angle PRT$, $\angle TSP$, $\angle QST$, and $\angle TRQ$ are all equal. We're halfway done!

Now, consider ΔTPR and ΔTQR. We have already determined that the lengths of line segments *PR* and *QR* are equal. It is trivial that line

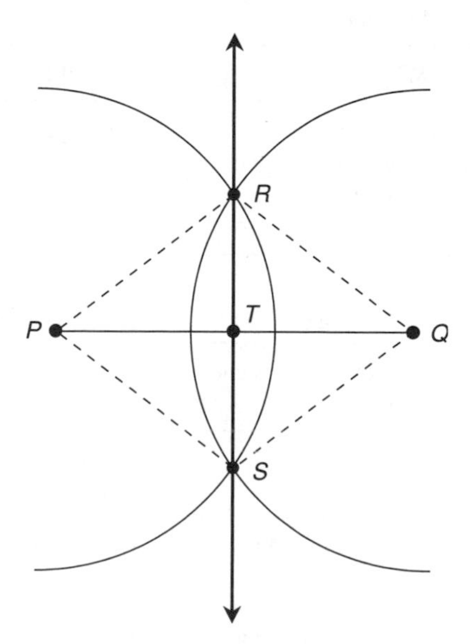

Fig. 7-8. Illustration for Problems 7-4 and 7-5.

segment *RT* has the same length as itself. We have determined that the measures of $\angle PRT$ and $\angle TRQ$ are equal. Thus, by the SAS axiom, ΔTPR is inversely congruent to ΔTQR. It follows from the definition of inverse congruence that line segments *TP* and *TQ* have the same length, and therefore, that line *RS* bisects line segment *PQ* at point *T*. Table 7-1 is an S/R version of this proof.

The above paragraph does not portray the only way to do the second half of this proof. There is at least one other way. Can you figure it out?

Table 7-1. An S/R version of the proof demonstrated in Solution 7-4.

Statements	**Reasons**
The lengths of line segments *RP*, *PS*, *SQ*, and *QR* are equal.	They are all radii of arcs generated with the compass set for the same span.
ΔSRP and ΔRSQ are isosceles.	We can conclude this directly from the previous statement.
In ΔSRP and ΔRSQ, corresponding pairs of sides have equal lengths going in the same direction.	This follows from what we already know, along with the fact that the triangles have line segment *RS* in common.
$\Delta SRP \equiv \Delta RSQ$.	This follows from the SSS axiom.
In ΔSRP and ΔRSQ, corresponding pairs of angles have equal measures going in the same direction.	You'll get a chance to fill this in later.
$m\angle PRT = m\angle TSP = m\angle QST = m\angle TRQ$.	This follows from the above, and the fact that ΔSRP and ΔRSQ are isosceles.
Line segment *RT* has the same length as itself.	This is trivial!
$\Delta TPR \equiv\!- \Delta TQR$.	This follows from facts we have already established, along with the SAS axiom.
Line segments *TP* and *TQ* are equally long.	This follows from the definition of inverse congruence.
Line *RS* bisects line segment *PQ* at point *T*.	This is evident from the geometry of the of situation.

PROBLEM 7-5

Prove that in the line-segment bisection method shown in Figs. 7-7 and 7-8, $\angle QTR$ and $\angle RTP$ are right angles, and that as a result of this fact, the process depicted in Fig. 7-7 provides a method of constructing a right angle.

SOLUTION 7-5

Refer again to Fig. 7-8. From Solution 7-4, we know that ΔTPR is inversely congruent to ΔTQR. According to the definition of inverse congruence, the corresponding angles $\angle QTR$ and $\angle RTP$ have equal measure. It is evident from the geometry of the situation that $\angle QTR$ and $\angle RTP$ are supplementary angles. That means their measures add up to a straight angle, which by definition has a measure of 180°. It follows from basic algebra that $\angle QTR$ and $\angle RTP$ both have measures of 90°, and are therefore both right angles. Thus, the process depicted in

Table 7-2. An S/R version of the proof demonstrated in Solution 7-5.

Statements	**Reasons**
$\Delta TPR \equiv - \Delta TQR$.	This was determined in Solution 7-4.
$m\angle QTR = m\angle RTP$.	This follows from the definition of inverse congruence, and from the fact that $\angle QTR$ and $\angle RTP$ are correspon-ding angles.
$\angle QTR$ and $\angle RTP$ are supplementary angles.	This is evident from the geometry of the situation.
$m\angle QTR + m\angle RTP = 180°$.	This follows from the definition of supplementary angles.
$m\angle QTR = 90°$ and $m\angle RTP = 90°$.	This follows from basic algebra and the statements in the second and fourth lines of this table.
$\angle QTR$ and $\angle RTP$ are both right angles.	This follows from the definition of right angle.
The process depicted in Fig. 7-7 provides a way of constructing a right angle.	This is true because $\angle QTR$ and $\angle RTP$ both arise during the process!

Fig. 7-7 provides a way of constructing a right angle. Table 7-2 is an S/R version of this proof.

BISECTING AN ANGLE

Fig. 7-9 illustrates one method that can be used to bisect an angle, that is, to divide it in half. First, suppose two rays intersect at a point P, as shown at A. Set down the non-marking tip of the compass on point P, and construct an arc from one ray to the other. Call the two points where the arc intersects the rays point R and point Q (Fig. 7-9B). We can now call the angle in question $\angle QPR$, where points R and Q are equidistant from point P.

Now, place the non-marking tip of the compass on point Q, increase its span somewhat from the setting used to generate arc QR, and construct a new arc. Next, without changing the span of the compass, set its non-marking tip down on point R and construct an arc that intersects the arc centered on point Q. (If the arc centered on point Q isn't long enough, go back and make it longer. You can make it a full circle if you want.) Let S be the point at which the two arcs intersect (Fig. 7-9C). Finally, construct ray PS, as shown at D. This ray bisects $\angle QPR$. This means that $m\angle QPS = m\angle SPR$, and also that $m\angle QPS + m\angle SPR = m\angle QPR$.

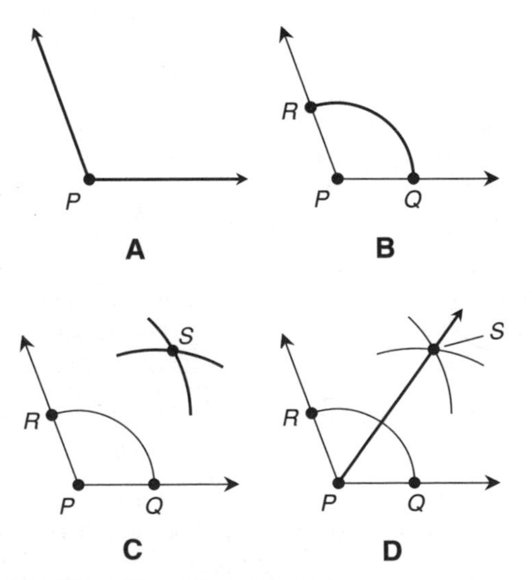

Fig. 7-9. Bisection of an angle.

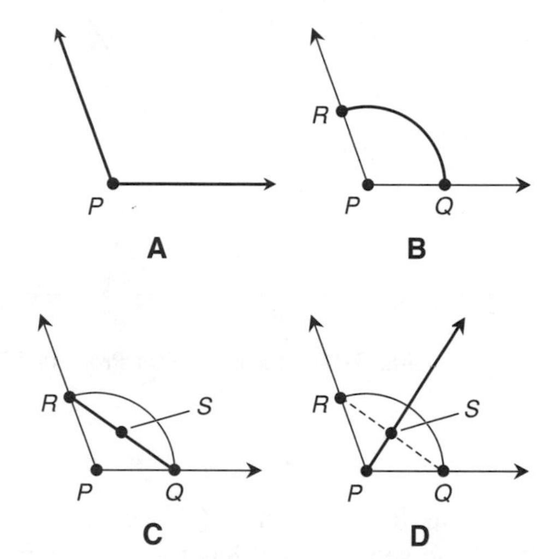

Fig. 7-10. Illustration for Problem 7-6.

PROBLEM 7-6

Find another way to bisect an angle.

SOLUTION 7-6

Refer to Fig. 7-10. The process starts in the same way as described above. Two rays intersect at point *P*, as shown in drawing A. Set down the non-marking tip of the compass on point *P*, and construct an arc from one ray to the other to get points *R* and *Q* (Fig. 7-10B) defining $\angle QPR$, where points *R* and *Q* are equidistant from point *P*.

Construct line segment *RQ*. Then bisect it, according to the procedure for bisecting line segments described earlier in this chapter. Call the midpoint of the line segment point *S*, as shown in Fig. 7-10C. Finally, construct ray *PS* (Fig. 7-9D). This ray bisects $\angle QPR$.

PROBLEM 7-7

Prove that the angle bisection method described in Solution 7-6 is valid according to the rules of construction and the principles of geometry.

SOLUTION 7-7

You may wish to expand and add some more detail to Fig. 7-10D, obtaining a drawing that looks like Fig. 7-11. Line segment *SR* has the

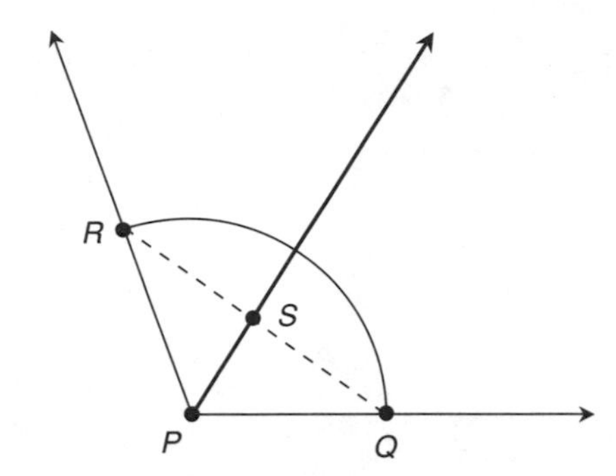

Fig. 7-11. Illustration for Problem 7-7.

same length as line segment *SQ*, because we have bisected line segment *RQ*. Line segment *RP* has the same length as line segment *QP*, because we have constructed them both from the same arc centered at point *P*. Line segment *PS* has the same length as line segment *PS*; this is trivial. From the geometry of this situation, it is evident that ΔSRP and ΔSQP have corresponding pairs of sides that are of equal lengths as you proceed around the triangles in opposite directions. From the SSS axiom and the definition of inverse congruence, we can conclude that ΔSRP and ΔSQP are inversely congruent. Therefore, corresponding pairs of angles (angles opposite corresponding sides), as we proceed around the triangles in opposite directions, have equal measure. (You'll get a chance to supply the reason for this in Quiz Question 2.) This means that the measure of $\angle SPR$ is equal to the measure of $\angle QPS$, because they constitute a pair of corresponding angles in ΔSRP and ΔSQP. The sum of the measures of these two angles is equal to the measure of $\angle QPR$; this is evident from the geometry of the situation. Therefore, according to the definition of angle bisection, ray *PS* bisects $\angle QPR$. Table 7-3 is an S/R version of this proof.

The Theorem of Pythagoras

One of the most famous mathematical facts ever proved is known as the *Theorem of Pythagoras*, also called the *Pythagorean Theorem*. It has been

known for thousands of years. Stated in words, it goes like this: "The square of the length of the longest side (*hypotenuse*) of a right triangle is equal to the sum of the squares of the lengths of the two shorter sides, if all the lengths are expressed in the same units." There is some debate as to who proved this theorem first, but many proofs have been done since ancient times. Here is one example of how this theorem can be proved.

Table 7-3. An S/R version of the proof demonstrated in Solution 7-7.

Statements	**Reasons**
Line segment *SR* has the same length as line segment *SQ*.	We have bisected line segment *RQ*, and the midpoint is *S*.
Line segment *RP* has the same length as line segment *QP*.	We have constructed them both from from the same arc centered at point *P*.
Line segment *PS* has the same length as itself.	This is trivial!
In ΔSRP and ΔSQP, corresponding pairs of sides have equal lengths as we go around them in opposite directions.	This is evident from the geometry of the situation.
$\Delta SRP \equiv\!- \Delta SQP$.	This follows from the SSS axiom and the definition of inverse congruence.
In ΔSRP and ΔSQP, corresponding pairs of angles have equal measures going in opposite directions.	You'll get a chance to fill this in later.
$m\angle SPR = m\angle QPS$.	These two angles constitute a pair of corresponding angles in ΔSRP and ΔSQP.
$m\angle SPR + m\angle QPS = m\angle QPR$.	This is evident from the geometry of the situation.
The ray *PS* bisects angle $\angle QPR$.	This follows from the definition of bisection.

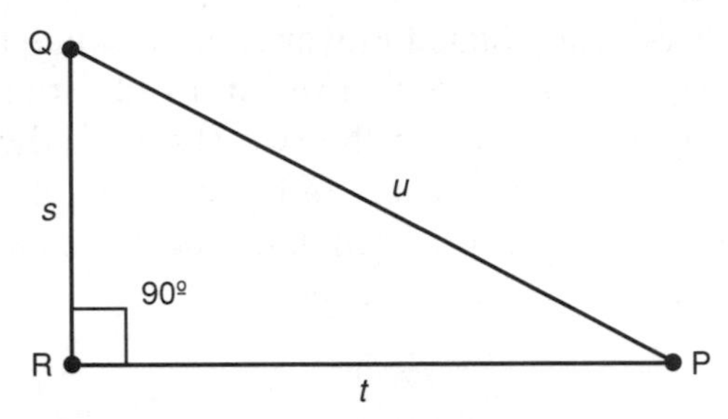

Fig. 7-12. Illustration of a right triangle for the statement of the Theorem of Pythagoras. It is always true that $s^2 + t^2 = u^2$.

THE PROPOSITION

Suppose we have a right triangle defined by points P, Q, and R whose sides have lengths s, t, and u respectively. Let u be the length of the hypotenuse (Fig. 7-12). Then the following equation is always true:

$$u^2 = s^2 + t^2$$

THE PROOF

Fig. 7-13 is a visual aid that, while not strictly necessary for the proof, makes things a lot easier to explain! To begin, imagine four directly congruent right triangles, each of whose sides have lengths of s, t, and u units, with the hypotenuses

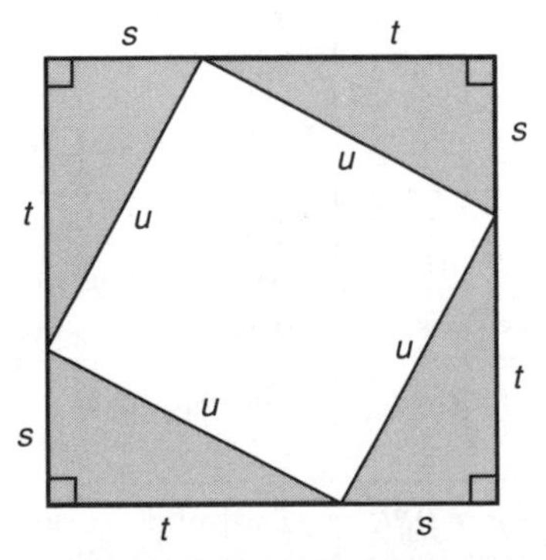

Fig. 7-13. Proof of the Theorem of Pythagoras. Four right triangles, all congruent to one another, are positioned to form a square within a square.

all measuring u units. Suppose they are arranged as shown in Fig. 7-13, so they form a large square whose sides each have length $s + t$ units. The area of this large square can be found by squaring the length of the side. (That is a well-known formula from geometry.) If we call the area of the large square A_L, then:

$$\begin{aligned} A_L &= (s + t)^2 \\ &= s^2 + 2st + t^2 \end{aligned}$$

where A_L is expressed in *square units* or *units squared.*

Look at the unshaded region inside the large square. It, too, is a square, measuring u units on each side. (Do you wonder how we know it is a square? You *should* be asking this question right now! You can prove that the unshaded region is a square by showing that each of its interior angles is a right angle. Consider this an "extra credit" exercise.) The area of the unshaded square, A_U, is equal u^2 square units. That is:

$$A_U = u^2$$

Consider the four right triangles. Each of them has an area, call it A_T, that is equal to half the height times the length of the base. (This is another well-known formula from geometry.) If we call s the height and t the length of the base for each right triangle, then:

$$A_T = st/2$$

There are four of these triangles, and the sum of their areas, call it A_S, indicated by the shaded regions in Fig. 7-13, is equal to:

$$\begin{aligned} A_S &= 4A_T \\ &= 4(st/2) \\ &= 2st \end{aligned}$$

It is apparent from the figure that $A_S + A_U = A_L$. That is, the sum of the shaded and unshaded regions is equal to the area of the large square. Substituting for each of these areas, we get the following result:

$$\begin{aligned} A_S + A_U &= A_L \\ 2st + u^2 &= s^2 + 2st + t^2 \end{aligned}$$

Subtracting the quantity $2st$ from each side of the second equation above gives us the formula for the Pythagorean Theorem:

$$u^2 = s^2 + t^2$$

The Square Root of 2

It is possible to construct a line segment whose length is, in theory, exactly equal to the positive square root of 2. (There's also a negative square root of 2 that is equal to −1 times the positive one, but we're concerned only with the positive square root of 2. Let's call it $2^{1/2}$, or 2 to the ½ power, for short.) Despite the fact that we can "build" it geometrically, the value of $2^{1/2}$ cannot be represented in arithmetic as a ratio of integers in lowest terms.

THE FIRST PROPOSITION

Let *PQ* be a line segment. Let *RS* be a line that has been constructed according to the process described earlier and illustrated by Fig. 7-7, so that line *RS* bisects line segment *PQ*. Let *T* be the point at which line *RS* intersects line segment *PQ*. According to Solution 7-5 derived earlier, ∠*QTR* is a right angle. Suppose we extend line segment *TQ* so it becomes a ray of indefinite length. Consider rays *TR* and *TQ*, as shown in Fig. 7-14A. Set a drafting compass for a span equal to

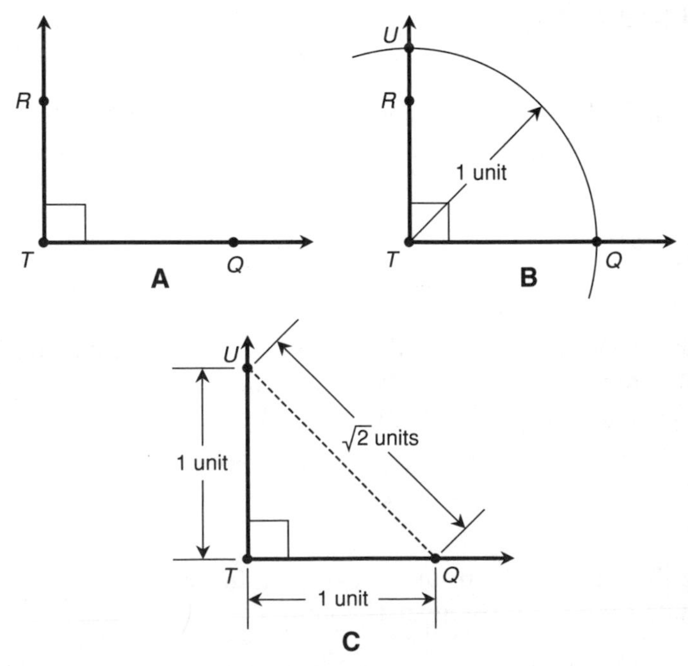

Fig. 7-14. Construction of a line segment having a length equal to the square root of 2 units.

the length of line segment TQ. Let's define this span as a distance of 1 unit. Using the compass, construct an arc having a radius equal to this span, centered at point T, and passing through rays TR and TQ (Fig. 7-14B). Let U be the point at which the arc intersects ray TR. Construct line segment UQ, as shown in Fig. 7-14C. Then the length of line segment UQ is equal to $2^{1/2}$ units.

PROOF OF THE FIRST PROPOSITION

Note that $\angle QTU$ is a right angle, because we constructed it that way. Therefore, by definition, ΔQTU is a right triangle. Let r be the length of line segment QT. Then r is equal to 1 unit, because we have defined it that way using the compass. Let q be the length of line segment TU. Then q is equal to 1 unit. (You'll get a chance to provide a reason for this in Quiz Question 3.) Let t be the length of line segment UQ. It is evident from the geometry of this situation that line segment UQ is the hypotenuse of ΔQTU. According to the Pythagorean Theorem, we have this formula:

$$t^2 = r^2 + q^2$$

By substitution and some basic arithmetic and algebra, we obtain this:

$$t^2 = 1^2 + 1^2$$
$$t^2 = 1 + 1$$
$$t^2 = 2$$
$$t = 2^{1/2}$$

Remember that t is defined as the length of line segment UQ. From the above derivation, it follows that the length of line segment UQ is equal to $2^{1/2}$ units. Table 7-4 is an S/R version of this proof.

THREE DEFINITIONS

Before we continue, let's review three definitions. You will recognize these from basic arithmetic. Here they are:

- A number n is an *even integer* if and only if $n/2$ is an integer.
- A number m is an *odd integer* if and only if $(m + 1)/2$ is an integer.
- A quotient of two integers a/b is a *ratio of integers in lowest terms* if and only if there does not exist any positive integer c such that a/c and b/c are both integers.

Table 7-4. An S/R version of the proof that the construction shown in Fig. 7-14, and described in the text, yields a line segment with a length equal to the square root of 2 (or $2^{1/2}$) units.

Statements	Reasons
$\angle QTU$ is a right angle.	We constructed it that way!
ΔQTU is a right triangle.	This follows from the definition of right triangle.
Let r be the length of line segment QT.	We have to call it something!
$r = 1$.	We defined the length of line segment QT as 1 unit, using the compass.
Let q be the length of line segment TU.	We have to call it something!
$q = 1$.	You'll get a chance to fill this in later.
Let t be the length of line segment UQ.	We have to call it something!
Line segment UQ is the hypotenuse of ΔQTU.	This is evident from the geometry of the situation.
$t^2 = r^2 + q^2$.	This is true according to the Theorem of Pythagoras.
$t^2 = 1^2 + 1^2$.	This is the result of substituting 1 for the values of r and q.
$t = 2^{1/2}$.	This is the result of solving the equation in the previous line for t.
The length of line segment UQ is equal to $2^{1/2}$ units.	This follows from the fact that we have defined t as the length of line segment UQ.

ODD-TIMES-ODD THEOREM

It will also help us to state a well-known fact from arithmetic. We won't go through its proof here, but you can prove it as an exercise if you want! Let's call it the *odd-times-odd theorem*. It states that the product of two odd integers is always an odd integer.

THE SECOND PROPOSITION

The value of $2^{1/2}$ cannot be represented as a ratio of integers in lowest terms.

PROOF OF THE SECOND PROPOSITION

Whenever we are confronted with the task of performing an "impossibility proof" or a "negativity proof," the situation suggests that we ought to try *reductio ad absurdum*. Let's use it here.

Suppose that the above proposition is false, and that the number $2^{1/2}$ can be represented as the ratio of two integers in lowest terms. Call those integers p and q. In order for the ratio to be defined, q must not be equal to 0. Here is what we claim:

$$2^{1/2} = p/q$$

Squaring both sides, we get:

$$(2^{1/2})^2 = (p/q)^2$$

This can be rewritten, using the rules of basic algebra, to obtain:

$$2 = p^2/q^2$$

Multiplying each side of the above equation by q^2 gives us this:

$$2q^2 = p^2$$

Dividing each side by 2, we get this:

$$q^2 = p^2/2$$

We stipulated that q is an integer; therefore q^2 is an integer. We also stipulated that p is an integer; therefore p^2 is an integer. (You'll get a chance to provide the reason for these two facts in Quiz Question 4.) The above equation therefore tells us that $p^2/2$ is an integer. By definition, then, p^2, which is equal to $p \times p$, is an even integer. It follows that p is an even integer; the odd-times-odd theorem guarantees that if p were odd, then p^2, or $p \times p$, would have to be odd. The fact that p is an even integer means, by definition, that $p/2$ is an integer. Let's call that integer t. Thus:

$$p/2 = t$$

Multiplying each side by 2 gives us the following:

$$p = 2t$$

Substituting $2t$ for p in the equation $2q^2 = p^2$ from above, we get:

$$2q^2 = (2t)^2$$

That can be simplified to:

$$2q^2 = 4t^2$$

Dividing each side by 4, we obtain:

$$q^2/2 = t^2$$

We know that t is an integer, so t^2 is an integer. (You'll get a chance to provide the reason for this in the chapter-ending quiz.) Thus $q^2/2$ is an integer. By definition, then, q^2 is even. It follows that q is an even integer; the odd-times-odd theorem guarantees that if q were odd, then q^2, or $q \times q$, would have to be odd. Therefore, by definition, $q/2$ is an integer.

Have patience! We're getting there!

All the way back at the beginning of this proof, we claimed that $2^{1/2}$ is equal to p/q, where p/q is a ratio of integers in lowest terms. We have shown that $p/2$ and $q/2$ are both integers. But then p/q, while a ratio of integers, is not in lowest terms, because $(p/2)/(q/2)$ is a ratio of integers equal to p/q! This contradicts our original assertion. Invoking *reductio ad absurdum*, we must reject our initial assumption, proving that the value of $2^{1/2}$ cannot be represented as a ratio of integers in lowest terms. Table 7-5 is an S/R version of this proof.

Table 7-5. An S/R version of the proof that the value of $2^{1/2}$ cannot be represented as a ratio of integers in lowest terms.

Statements	**Reasons**
Assume $2^{1/2}$ can be represented as a ratio of two integers, p and q, in lowest terms.	This is the initial assumption, from which we will derive a contradiction and then apply *reductio ad absurdum*.
$2^{1/2} = p/q$.	This is merely a specification of the claim made above.
$(2^{1/2})^2 = (p/q)^2$.	Square both sides of the equation in the previous line.
$2 = p^2/q^2$.	Use basic algebra on the equation in the previous line.
$2q^2 = p^2$.	Multiply each side of the equation in the previous line by q^2.
$q^2 = p^2/2$.	Divide each side of the equation in the previous line by 2.
We know $q \in \mathbf{Z}$. Therefore $q^2 \in \mathbf{Z}$.	You'll get a chance to fill this in later.
We know $p \in \mathbf{Z}$. Therefore $p^2 \in \mathbf{Z}$.	You'll get a chance to fill this in later.

Table 7-5. *(continued)*

Statements	**Reasons**
$p^2/2 \in \mathbf{Z}$.	We know this because $q^2 = p^2/2$, and $q^2 \in \mathbf{Z}$.
p^2 is an even integer.	This follows from the definition of even integer.
p is an even integer.	According to the odd-times-odd theorem, if p were odd, then $p \times p$ would be odd.
$p/2 \in \mathbf{Z}$.	This follows from the definition of even integer.
Let $p/2 = t$, where $t \in \mathbf{Z}$.	This will make things simpler!
$p = 2t$.	Multiply each side of the equation in the previous line by 2.
$2q^2 = (2t)^2$.	Substitute $2t$ for p in the equation $2q^2 = p^2$ from above.
$2q^2 = 4t^2$.	Simplify the right-hand side of the equation in the previous line.
$q^2/2 = t^2$.	Divide each side of the equation in the previous line by 4.
We know $t \in \mathbf{Z}$. Therefore $t^2 \in \mathbf{Z}$.	You'll get a chance to fill this in later.
$q^2/2 \in \mathbf{Z}$.	This follows from the previous two lines.
q^2 is an even integer.	This follows from the definition of even integer.
q is an even integer.	According to the odd-times-odd theorem, if q were odd, then $q \times q$ would be odd.
$q/2 \in \mathbf{Z}$.	This follows from the definition of even integer.
The quotient p/q is a ratio of integers in lowest terms.	This is part of the assertion we made to start this proof.
$p/2 \in \mathbf{Z}$ and $q/2 \in \mathbf{Z}$.	We have proven both of these facts.
$(p/2)/(q/2)$ is a ratio of integers.	This follows from the statement immediately above this line.
The ratio p/q is not given in lowest terms.	This follows from the statement immediately above this line, and the fact that $(p/2)/(q/2) = p/q$.
We have a contradiction!	The line immediately above this one is contrary to our original assertion.
$2^{1/2}$ is not a ratio of integers in lowest terms.	Invoke *reductio ad absurdum*.

The Greatest Common Divisor

Here is a well-known fact of arithmetic that involves the positive integers. It can be used to prove a lot of interesting things. Let's state it and prove it!

THE GCD THEOREM

Let p and q be positive integers. Then there exists a unique largest positive integer g such that p/g and q/g are both positive integers. The positive integer g is called the *greatest common divisor* (GCD) for p and q. Sometimes it is also called the *greatest common factor* (GCF) or the *highest common factor* (HCF). Let's call this the *GCD theorem*. In order to prove it, we'll need a well-known axiom.

THE TRICHOTOMY AXIOM

For all real numbers p and q, one and only one of three conditions holds: (1) p is smaller than q, or (2) p is larger than q, or (3) p is equal to q. That is, exactly one of the following is true:

$$p < q$$
$$p > q$$
$$p = q$$

This is also known as the *principle of trichotomy*, or simply as *the trichotomy*.

PROOF OF THE GCD THEOREM

Let's denote the set of positive integers by the uppercase, bold, italic English letter $\mathbf{Z}$ with a plus-sign subscript ($\mathbf{Z}_+$). Imagine two positive integers p and q. The number 1 is a common divisor for all positive integers. It's possible that 1 is the only positive integer that divides both p and q without leaving remainders. In other words, it is possible that there is no positive integer g, other than $g = 1$, such that p/g and q/g are both elements of $\mathbf{Z}_+$. In that case, 1 is the GCD, and the proof is over! But it is also possible that 1 is not the only positive integer that divides both p and q without remainders. (In Quiz Question 5, you will get a chance to identify a pair of positive integers p and q that have this characteristic.)

Consider three cases according to the trichotomy as the axiom applies to common divisors.

Case 1: Suppose that p is smaller than q. Then no positive integer n larger than p divides both p and q without remainders, because if such an n did exist, then p/n would be between 0 and 1, and would not be a positive integer. It follows that the GCD must exist, and it must be somewhere between 1 and p inclusive; that is, $1 \leq g \leq p$.

Case 2: Suppose that p is larger than q. Then no positive integer n larger than q divides both p and q without remainders, because if such an n did exist, then q/n would be between 0 and 1, and would not be a positive integer. It follows that the GCD must exist, and it must be somewhere between 1 and q inclusive; that is, $1 \leq g \leq q$.

Case 3: Suppose $p = q$. Then both p and q divide each other without remainders. But no positive integer n larger than p or q can divide either of them without remainders. If that were the case, then p/n, which is the same as q/n, would have to be between 0 and 1, and this number would therefore not be a positive integer. It follows that the GCD must exist, and it must be equal to both p and q; that is, $g = p = q$.

We have shown that in all cases of the trichotomy, a GCD exists. It follows that there is a GCD for any pair of positive integers. Table 7-6 is an S/R version of this proof.

Table 7-6. An S/R version of the proof of the GCD theorem.

Statements	Reasons
Let $\mathbf{Z}_+$ denote the set of positive integers.	We have to symbolize it somehow!
Suppose that $p \in \mathbf{Z}_+$ and $q \in \mathbf{Z}_+$.	We need two nonspecific positive integers to do the proof.
If a GCD exists for p and q, let's call it g.	We have to call it something!
Suppose 1 is the only positive integer that divides p and q without remainders.	This is the first of two possibilities Let's see what it implies.
$g = 1$, and there is nothing further to prove.	This follows from the definition of GCD.

Table 7-6. *(continued)*

Statements	**Reasons**
Suppose 1 is not the only positive integer that divides p and q without remainders.	This is the second of two possibilities. Let's see what it implies.
One and only one of the following is true: $p < q$, or $p > q$, or $p = q$.	This follows from the trichotomy axiom because p and q, being positive integers, are real numbers as well.
Suppose that $p < q$.	This is case 1 of the trichotomy. Let's see what it implies.
There is no positive integer n larger than p such that $p/n \in \mathbf{Z}_+$ and $q/n \in \mathbf{Z}_+$.	If such an n exists, then $0 < p/n < 1$, and thus $p/n \notin \mathbf{Z}_+$.
$1 \leq g \leq p$.	We know 1 divides both p and q without remainders, and that no positive integer larger than p can do so. We are thus forced to this conclusion.
Suppose that $p > q$.	This is case 2 of the trichotomy. Let's see what it implies.
There is no natural number n larger than q such that $p/n \in \mathbf{Z}_+$ and $q/n \in \mathbf{Z}_+$.	If such an n exists, then $0 < q/n < 1$, and thus $q/n \notin \mathbf{Z}_+$.
$1 \leq g \leq q$.	We know 1 divides both p and q without remainders, and that no positive integer larger than q can do so. We are thus forced to this conclusion.
Suppose that $p = q$.	This is case 3 of the trichotomy. Let's see what it implies.
$p/q \in \mathbf{Z}_+$ and $q/p \in \mathbf{Z}_+$.	This is trivial because $p/q = q/p = 1$, and $1 \in \mathbf{Z}_+$.
There is no natural number n larger than p and q such that $p/n \in \mathbf{Z}_+$ or $q/n \in \mathbf{Z}_+$.	If such an n exists, then $0 < p/n < 1$ making $p/n \notin \mathbf{Z}_+$, and $0 < q/n < 1$ making $q/n \notin \mathbf{Z}_+$.
$g = p = q$.	We know that p and q divide each other without remainders, and that no number larger than them can do so. We are thus forced to this conclusion.
There exists a GCD for any pair of positive integers.	All the possibilities have been covered!

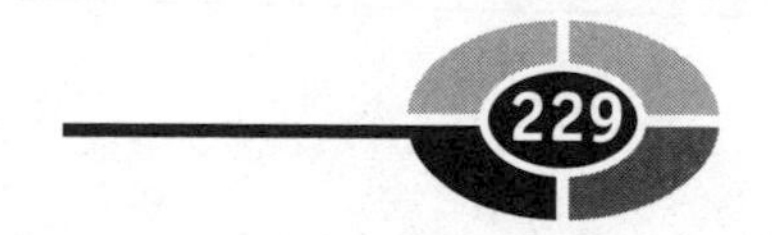

Prime Numbers

Euclid, known for his work in geometry, also proved important theorems about numbers. Here are a couple of his theorems that involve so-called *prime numbers*. Before we get started with these famous proofs, however, we need a famous axiom! We should also define what a prime number is, and what we should call the numbers that are not prime.

THE WELL-ORDERING AXIOM

Every non-empty set of positive integers contains a smallest element.

WHAT IS A PRIME?

Let n be a positive integer larger than 1. The number n is a prime number (also called a *prime*) if and only if, when n is divided by a positive integer k, the quotient n/k is an integer only when $k = 1$ or $k = n$. Stated another way, a prime number is a positive integer larger than 1 that is divisible by a positive integer without a remainder only when the divisor is equal to 1 or the number itself. The set of all prime numbers is sometimes denoted by the uppercase, bold, italic English letter ***P***.

WHAT IS A COMPOSITE?

The number n is a *composite number* (also called a *composite*) if and only if n is a positive integer, n is not equal to 1, and n is not a prime number. The set of all composite numbers is sometimes denoted by the uppercase, bold, italic English letter ***C***.

WHAT ABOUT 1?

The above definitions deal only with positive integers larger than 1. This is a matter of convention. The positive integer 1 is not considered prime, even though it is divisible without a remainder only when the divisor is equal to 1 or itself. But it is not considered composite. Whenever you hear about prime or composite numbers, then, remember that such numbers are always positive integers larger than or equal to 2.

THE PRIME-FACTOR THEOREM

Any composite number can be expressed as a product of primes. A good descriptive name for this is the *prime-factor theorem*. It is also known as the "weak version" of the *Fundamental Theorem of Arithmetic*.

PROOF OF THE PRIME-FACTOR THEOREM

Suppose there are composite numbers that are not products of primes. According to the well-ordering axiom, there is a smallest such number. Call it x. We know that x is larger than 1. (You'll get a chance to provide a reason for this in Quiz Question 6.) We also know that x is not prime. It follows that there are positive integers y and z, both larger than 1 and less than x, such that $x = yz$. (We know that y and z are both larger than 1 and less than x, because if y or z had to be equal to either 1 or x, then x would be prime; and if y or z were larger than x, then x/y or x/z would be between 0 and 1, and would thus not be a positive integer.) We now have four possible cases:

- Both y and z are prime. It follows that x is a product of primes, because $x = yz$, and y and z are both prime.
- The number y is prime, but z is composite. Because z is smaller than x, and x is the smallest composite that is not a product of primes, z must be a product of primes. Because $x = yz$ and y is prime, it follows that x is a product of primes.
- The number z is prime, but y is composite. Because y is smaller than x, and x is the smallest composite that is not a product of primes, y must be a product of primes. Because $x = yz$ and z is prime, it follows that x is a product of primes.
- Both y and z are composite. They are both less than x, and we know that x is the smallest composite number that is not a product of primes. Therefore, y and z must both be products of primes. Because $x = yz$, then, x is also a product of primes.

All four of these cases contradict the assertion that x is not a product of primes. That means our original assertion is false. There are no composite numbers that are not products of primes. This can be more simply stated as the proposition we intended to prove: Any composite number can be expressed as a product of primes. Table 7-7 is an S/R version of this proof.

Table 7-7. An S/R version of the proof of the prime-factor theorem, also known as the "weak version" of the Fundamental Theorem of Arithmetic.

Statements	Reasons
Suppose there exist composite numbers that are not products of primes.	This is the initial assumption from which we will derive a contradiction, thus proving its negation, which is the proposition at hand.
There is a smallest composite number x that is not a product of primes.	This follows from the well-ordering axiom.
$x > 1$.	You'll get a chance to fill this in later.
$x \notin \mathbf{P}$.	We are told that x is composite, and by definition, this means x is not prime.
There exist positive integers y and z, such that $x = yz$.	This follows from the definition of prime number.
$1 < y < x$.	If y had to be equal to either 1 or x, then x would be prime; and if y were larger than x, then x/y would would be between 0 and 1, and would not be a positive integer.
$1 < z < x$.	If z had to be equal to either 1 or x, then x would be prime; and if z were larger than x, then x/z would would be between 0 and 1, and would not be a positive integer.
Suppose that $(y \in \mathbf{P})$ & $(z \in \mathbf{P})$.	This is the first of four possible cases.
The number x is a product of primes.	This follows from the facts that $x = yz$, $y \in \mathbf{P}$, and $z \in \mathbf{P}$.
Suppose that $(y \in \mathbf{P})$ & $(z \in \mathbf{C})$.	This is the second of four possible cases.
The number z is a product of primes.	This follows from the facts that $z < x$ and x is the smallest composite that is not a product of primes.
The number x is a product of primes.	This follows from the facts that $x = yz$ and $y \in \mathbf{P}$.
Suppose that $(y \in \mathbf{C})$ & $(z \in \mathbf{P})$.	This is the third of four possible cases.
The number y is a product of primes.	This follows from the facts that $y < x$ and x is the smallest composite that is not a product of primes.

Table 7-7. *(continued)*

Statements	Reasons
The number x is a product of primes.	This follows from the facts that $x = yz$ and $z \in \boldsymbol{P}$.
Suppose that $(y \in \boldsymbol{C})$ & $(z \in \boldsymbol{C})$.	This is the fourth of four possible cases.
$(y < x)$ & $(z < x)$.	We determined this earlier.
The number x is the smallest composite that is not a product of primes.	We determined this earlier.
The numbers y and z are both products of primes.	This follows from the previous two steps.
The number x is a product of primes.	This follows from the previous step and the fact that $x = yz$.
All four of the above cases result in contradictions.	In each case, we determine that x is a product of primes, but we determined earlier that x is not a product of primes.
There exist no composite numbers that are not products of primes.	We are forced to conclude this because it is the negation of our original assertion.
Any composite number is a product of primes.	This is a simpler way of expressing the previous statement.

THE PRIME-FACTOR COROLLARY

Every composite number can be divided by at least one prime without leaving a remainder. That is, if n is a composite number, then there exists at least one prime number p such that n/p is a positive integer.

THE PROOF

Let n be a composite number. Because n is composite, n is expressible as a product of primes. Suppose those primes are p_1, p_2, p_3, . . . , and p_m, such that:

$$n = p_1 \times p_2 \times p_3 \times \ldots \times p_m$$

where m is a positive integer larger than 1. Now suppose we divide this product of primes (all of which are integers) by p_1. Then we have the following:

$$n/p_1 = (p_1 \times p_2 \times p_3 \times \ldots \times p_m) / p_1$$
$$= p_2 \times p_3 \times \ldots \times p_m$$

This is a product of primes (all of which are integers). Call this product k. We know that k is an integer. (You'll get a chance to supply the reason for this in Quiz Question 7). That means p_1 divides n without a remainder, and therefore, that n can be divided by at least one prime without a remainder. Table 7-8 is an S/R version of this proof.

Table 7-8. An S/R version of the proof of the prime-factor corollary.

Statements	**Reasons**
Suppose $n \in \mathbf{C}$.	This is our starting point.
The number n is expressible as a product of primes.	This follows from the prime factor theorem.
$n = p_1 \times p_2 \times p_3 \times \ldots \times p_m$, where $p_1, p_2, p_3, \ldots, p_m \in \mathbf{P}$, $m \in \mathbf{Z}_+$, and $m > 1$.	This is a restatement of the fact that n is expressible as a product of primes.
$n/p_1 = p_2 \times p_3 \times \ldots \times p_m$	This is the result of dividing each side of the preceding equation by the prime p_1.
$p_2 \times p_3 \times \ldots \times p_m$ is a product of primes.	We know this because $p_1, p_2, p_3, \ldots, p_m \in \mathbf{P}$.
Let $p_2 \times p_3 \times \ldots \times p_m = k$.	We simply rename the product.
The number k is an integer.	You'll get a chance to fill this in later.
The prime p_1 divides n with no remainder.	This follows by algebra from the preceding five steps.
The number n can be divided by at least one prime without a remainder.	We've found a prime that does it, namely, p_1.

THE NO-LARGEST-PRIME THEOREM

There exists no largest prime number.

THE PROOF

Suppose there is a largest prime number. Call it p. Let S_p be the set of all primes less than or equal to p, as follows:

$$S_p = \{2, 3, 5, 7, 11, 13, \ldots, p\}$$

Multiply together all of the elements of S_p, and then add 1. Let the result be called z. Then:

$$z = (2 \times 3 \times 5 \times 7 \times 11 \times 13 \times \ldots \times p) + 1$$

Clearly, $z > p$, because z is 1 larger than at least one positive-integer multiple of p. From the product-of-integers and sum-of-integers axioms, we know that z is an integer, because z is a product of integers (the elements of S_p) with an integer added (1). For every prime number k in S_p, the quotient z/k has a remainder of 1. This is true because z/k is equal to a product of primes, plus 1. It follows that no element of S_p divides z without a remainder. We now have two possible cases:

- The number z is prime. It follows that p is not the largest prime, because $z > p$.
- The number z is composite. It follows that z is divisible without a remainder by at least one prime. (You'll get a chance to provide the reason for this in Quiz Question 8.) Let q be such a prime. We know that q is not an element of S_p, because no element of S_p divides z without a remainder. Therefore $q > p$, so p is not the largest prime.

Both of these cases contradict the assertion that p is the largest prime. That means our original assumption, that a largest prime exists, is false. There is no largest prime number. Table 7-9 is an S/R version of this proof.

PROBLEM 7-8

Prove that there are infinitely many prime numbers.

SOLUTION 7-8

Reductio ad absurdum suggests itself here, yet again. (You'll get a chance to provide the reason for this in Quiz Question 9.) To begin, assume that the number of primes is finite. Let the set of primes be

Table 7-9. An S/R version of the proof of the no-largest-prime theorem.

Statements	Reasons
Suppose there is a largest prime number.	This is our initial assumption, from which we will derive a contradiction.
Let p be the largest prime.	We're simply naming the number.
Let $\mathbf{S}_p$ be the set of all primes smaller than or equal to p.	We're simply naming the set.
Let z be the product of all the elements of $\mathbf{S}_p$, plus 1.	We're simply naming the quantity.
$z > p$	This follows from the fact that z is 1 larger than at least one positive-integer multiple of p.
$z \in \mathbf{Z}$	This follows from the product-of-integers axiom and the sum-of-integers axiom.
Suppose $k \in \mathbf{S}_p$.	There exists at least one such; let's just investigate its properties.
The quotient z/k always has a remainder of 1, no matter which k we choose from $\mathbf{S}_p$.	The quotient z/k is equal to a product of primes, plus 1.
No element of $\mathbf{S}_p$ divides z without a remainder.	This is a rewording of the previous statement using a double negative.
Suppose $z \in \mathbf{P}$.	This is the first of two possible cases.
The number p is not the largest prime.	This follows from the fact that $z > p$, which we have established.
Suppose $z \in \mathbf{C}$.	This is the second of two possible cases.
The number z is divisible without a remainder by at least one prime.	You'll get a chance to fill this in later.
Let q be a prime that divides z without a remainder.	We're simply identifying one such, and giving it a name.
$q \notin \mathbf{S}_p$	We have established that no element of $\mathbf{S}_p$ divides z without a remainder, but q does.
$q > p$	The prime q can't be less than or equal to p, because then it would be an element of $\mathbf{S}_p$!

Table 7-9 *(continued)*

Statements	**Reasons**
The number p is not the largest prime.	We just found one bigger, namely, q.
Both of the above cases result in contradictions.	In each case, we determine that p is not the largest prime, but we assumed originally that it is.
There is no largest prime.	We are forced to conclude this because it is the negation of our original assertion.

denoted by $\boldsymbol{P} = \{p_1, p_2, p_3, \ldots, p_m\}$, such that p_1 is less than p_2, which in turn is less than p_3, and so on up to p_m. That means the largest prime is p_m, because it is the last element in a finite, ascending sequence of numbers. But that is a contradiction. (You'll get a chance to provide the reason for this in Quiz Question 10.) Therefore, we must conclude that it is not true that the number of primes is finite. In other words, there are infinitely many primes. Table 7-10 is an S/R version of this proof.

Table 7-10. An S/R version of the proof demonstrated in Solution 7-8.

Statements	**Reasons**
Assume that the number of primes is finite.	We will derive a contradiction from this, thereby proving its negation.
Let the set of primes be denoted by $\boldsymbol{P} = \{p_1, p_2, p_3, \ldots, p_m\}$, such that $p1 < p2 < p3 < \ldots < p_m$.	We simply list the primes in ascending order.
The largest prime number is p_m.	It is the last element in a finite, ascending sequence of numbers.
The preceding statement is a contradiction.	You'll get a chance to fill this in later.
It is not true that the number of primes is finite.	This is the negation of our original assumption.
There are infinitely many prime numbers.	This is a rewording of the previous statement.

Quiz

This is an "open book" quiz. You may refer to the text in this chapter. A good score is 8 correct. Answers are in the back of the book.

1. In Solution 7-4 and Table 7-1, we claim that the measures of the corresponding pairs of angles, as we proceed around ΔSRP and triangle ΔRSQ in the same direction, are equal. What allows us to be certain of this?
   (a) The definition of isosceles triangle.
   (b) The SAA axiom.
   (c) The AAA axiom.
   (d) The definition of direct congruence.

2. In Solution 7-7 and Table 7-3, there is a statement to the effect that in ΔSRP and ΔSQP, corresponding pairs of angles have equal measures going in opposite directions. What allows us to make this claim?
   (a) The SAS axiom.
   (b) The definition of inverse congruence.
   (c) Euclid's original four postulates.
   (d) The definition of angle bisection.

3. In the first proof in the section called "The Square Root of 2," and also in Table 7-4, a statement is made that q is equal to 1 unit. We can make this claim because
   (a) q is the length of the hypotenuse of the triangle, and we have already assumed that this distance is equal to 1 unit.
   (b) q is equal the distance between points T and R, which was defined as equal to 1 unit of distance.
   (c) q is equal to the length of line segment TU, which was constructed with the compass set for a span equal to the length of line segment QT, a distance known to be 1 unit.
   (d) q is equal to the length of the arc made by the compass between points U and Q, centered on point T.

4. Refer back to the proof that $2^{1/2}$ cannot be expressed as a ratio of integers in lowest terms (the "proof of the second proposition") along with Table 7-5. It says the fact that q is an integer implies that q^2 is an integer, the fact that p is an integer implies that p^2 is an integer, and the fact that t is an integer implies that t^2 is an integer. What allows us to make these claims?

(a) The product-of-integers axiom from Chapter 6.
(b) The sum-of-integers axiom from Chapter 6.
(c) The definition of rational number from Chapter 6.
(d) The definition of set intersection from Chapter 6.

5. In the proof of the GCD theorem, it says that you'll get a chance to provide examples of two positive integers, p and q, such that 1 is not the only natural number that "goes cleanly into" them both. Which of the following pairs of positive integers meets that description?

   (a) $p = 7$ and $q = 21$
   (b) $p = 9$ and $q = 63$
   (c) $p = 19$ and $q = 95$
   (d) All of the above

6. In the proof of the prime-factor theorem, and in its S/R version (Table 7-7), it is stated that we know x is larger than 1. What allows us to make this claim?

   (a) The fact that x is a prime number, and the fact that all prime numbers are larger than 1, together imply that x is larger than 1.
   (b) The fact that x is a composite number, and the fact that all composite numbers are larger than 1, together imply that x is larger than 1.
   (c) The well-ordering axiom ensures that x is larger than 1 because it is the smallest positive integer.
   (d) We have assumed that the quotient x/x is equal to x, and 1 is the only natural number that has this property.

7. In the proof of the prime-factor corollary, and in its S/R version (Table 7-8), it is stated that we know k is an integer. How do we know this?

   (a) It follows from the quotient-of-fractions axiom in Chapter 6.
   (b) It follows from the product-of-integers axiom in Chapter 6.
   (c) It follows from the definition of integer in Chapter 6.
   (d) It follows from the definition of rational number in Chapter 6.

8. In the proof of the no-largest-prime theorem, and in its S/R version (Table 7-9), it is stated that the number z is divisible without a remainder by at least one prime. How do we know this?

   (a) It follows from the quotient-of-integers axiom.
   (b) It follows from the definition of prime number.
   (c) It follows from the definition of composite number.
   (d) It follows from the prime-factor corollary.

9. In the opening of Solution 7-8, it is stated that *reductio ad absurdum* suggests itself. What is the reason for this?
   (a) *Reductio ad absurdum* is the best course of action for any proof when we are not sure how to approach it; and if ever there was a difficult-looking problem, this is it!
   (b) *Reductio ad absurdum* suggests itself in "impossibility proofs" or "negativity proofs," and Problem 7-8 is a problem of this sort in disguise, because we are told to prove that a certain set is not finite.
   (c) *Reductio ad absurdum* is always the best course of action when we are confronted with a proof involving a finite set of natural numbers.
   (d) *Reductio ad absurdum* is always the best course of action when we are confronted with a proof involving the prime numbers.

10. In Solution 7-8, and in its S/R rendition (Table 7-10), it is stated that a contradiction occurs. What theorem or axiom is contradicted?
    (a) The prime-factor theorem.
    (b) The prime-factor corollary.
    (c) The no-largest-prime theorem.
    (d) The well-ordering axiom.

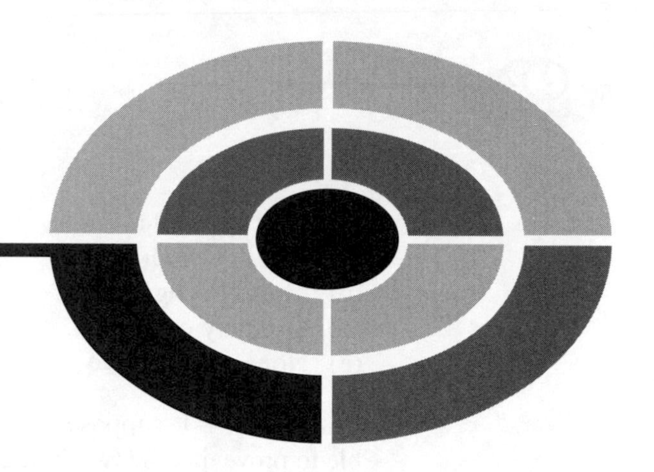

Test: Part Two

Do not refer to the text when taking this test. You may draw diagrams or use a calculator if necessary. A good score is at least 23 answers (75 percent or more) correct. Answers are in the back of the book. It's best to have a friend check your score the first time, so you won't memorize the answers if you want to take the test again.

1. Suppose you want to prove that a certain number x is a rational number. You can do this by showing that x
   (a) can be expressed as the product of two integers a and b, where $b \neq 0$.
   (b) can be expressed as the sum of two integers a and b, where $b \neq 0$.
   (c) can be expressed as the difference between two integers a and b, where $b \neq 0$.
   (d) can be expressed in the form a^b (a to the bth power), where a and b are integers and $b \neq 0$.
   (e) None of the above

2. In Fig. Test 2-1, suppose $a = c$, $b = d$, and $x = y$. These facts make it possible to prove that the two triangles, formed by the vertex points shown, are

(a) directly similar.
(b) inversely similar.
(c) directly congruent.
(d) inversely congruent.
(e) None of the above

3. In Fig. Test 2-1, suppose $a/c = b/d$, and $x = y$. These facts make it possible to prove that the two triangles, formed by the vertex points shown, are
   (a) directly similar.
   (b) inversely similar.
   (c) directly congruent.
   (d) inversely congruent.
   (e) None of the above

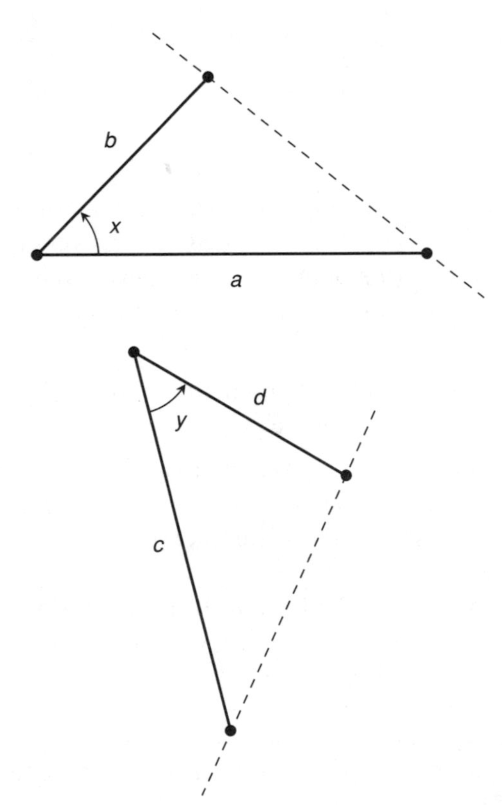

Fig. Test 2-1. Illustration for Part Two Test Questions 2 and 3.

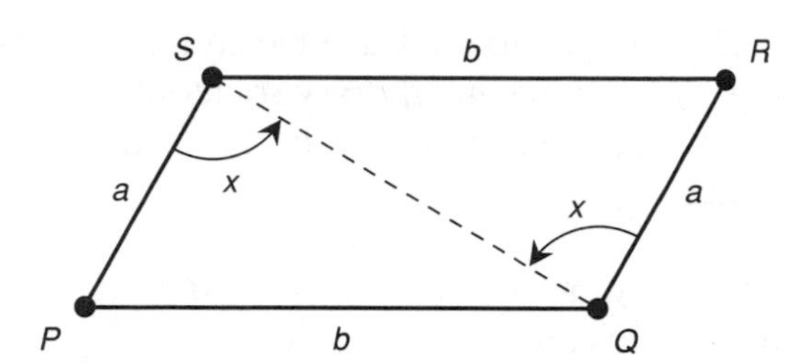

Fig. Test 2-2. Illustration for Part Two Test Questions 4 and 5.

4. In Fig. Test 2-2, suppose the figure *PQRS* has sides and angles of lengths and measures as shown. If you want to prove that ΔPQS is directly congruent to ΔRSQ, you
   (a) can do it easily using the side-side-side (SSS) axiom.
   (b) can do it easily using the side-angle-side (SAS) axiom.
   (c) can do it easily using either the SSS axiom or the SAS axiom.
   (d) can do it, but not using either the SSS axiom or the SAS axiom.
   (e) cannot do it, because it is not in general true.

5. In Fig. Test 2-2, suppose the figure *PQRS* has sides and angles of lengths and measures as shown. If you want to prove that ΔPQS is inversely congruent to ΔRSQ, you
   (a) can do it easily using the side-side-side (SSS) axiom.
   (b) can do it easily using the side-angle-side (SAS) axiom.
   (c) can do it easily using either the SSS axiom or the SAS axiom.
   (d) can do it, but not using either the SSS axiom or the SAS axiom.
   (e) cannot do it, because it is not in general true.

6. In a formal geometric construction using a compass and straight edge, which of the following actions (a), (b), (c), or (d) is impossible or not allowed?
   (a) Using a compass to draw a circle around a specific point, having an arbitrary (randomly chosen) radius.
   (b) Using a compass to draw a circle around a specific point, having a radius equal to the length of a specific line segment.
   (c) Using a compass to draw a circle around a specific point, having a radius equal to 5 centimeters.

(d) Using a compass to draw a circle around an arbitrary (randomly chosen) point, having an arbitrary (randomly chosen) radius.
(e) All of the above actions (a), (b), (c), and (d) are possible, and all are allowed.

7. In Euclidean geometry, three terms are defined only in an informal way. These three terms are
   (a) point, triangle, and sphere.
   (b) triangle, polynomial, and sphere.
   (c) line, angle, and circle.
   (d) plane, line, and point.
   (e) ray, sphere, and cube.

8. Table Test 2-1 is an S/R proof that any element in the intersection of two sets is also in their union. There are two blanks in the **Reasons** column. What words should go in the first blank (the one in the fourth line)?
   (a) "the definition of logical disjunction."
   (b) "the definition of set intersection."
   (c) "the definition of set union."
   (d) "the law of implication reversal."
   (e) "DeMorgan's law for conjunction."

9. What words should go in the second blank (the one in the sixth line) of Table Test 2-1?
   (a) "the definition of logical disjunction."
   (b) "the definition of set intersection."
   (c) "the definition of set union."
   (d) "the law of implication reversal."
   (e) "DeMorgan's law for conjunction."

10. Suppose we are told that there are two sets called G and H, neither one of them empty, such that $G \subset H$. From this, which of the following statements can be easily proved?
    (a) $(\forall x)\ (x \in G \Leftrightarrow x \in H)$
    (b) $(\forall x)\ (x \in H \Rightarrow x \in G)$
    (c) $(\forall x)\ (x \notin H \Rightarrow x \notin G)$
    (d) $(\exists x)\ (x \in G\ \&\ x \notin H)$
    (e) $(\forall x)\ (x \in G\ \&\ x \notin H)$

Table Test 2-1. An S/R proof that every element in the intersection of two sets is in the union of those two sets. This table goes with Part Two Test Questions 8 and 9.

Statements	**Reasons**
Suppose that X and Y are sets, each containing at least one element.	We will use these in the proof.
Let k be a constant.	We will use this in the proof.
Imagine that k is an element of the set $X \cap Y$.	This is our initial assumption.
From the above, we know that the constant k is an element of set X, and also that k is an element of set Y.	This follows from ________.
The constant k is an element of set X.	This follows from the definition of of logical conjunction; and anyhow, it's part of the preceding statement.
From the above, we know that k is an element of set X, or k is an element of set Y.	This follows from ________.
The constant k is an element of the set $X \cup Y$.	This follows from the definition of set union.

11. Suppose we have two triangles, and we want to prove that they are directly congruent. There are several ways we can do this. Which, if any, of the following approaches (a), (b), (c), or (d) is *not* sufficient to prove direct congruence for triangles?
    (a) Side-side-side.
    (b) Angle-angle-angle.
    (c) Side-angle-side.
    (d) Angle-side-angle.
    (e) Any of the above approaches is sufficient to prove direct congruence for triangles.

12. Table Test 2-2 is an S/R proof that
    (a) some integers are rational numbers.
    (b) all integers are rational numbers.

Table Test 2-2. An S/R proof. This table goes with Part Two Test Questions 12, 13, and 14.

Statements	**Reasons**
Suppose k is an integer.	We make this assumption to begin the proof.
When k is divided by 1, the quotient is equal to k.	This follows from the division-by-1 axiom.
The number 1 is an integer.	This follows from the definition of the set of integers.
The value of k is equal to the quotient of two integers, and the denominator is nonzero.	In the expression ________, the numerator and denominator are both integers, and 1 is not equal to 0.
The number k is an element of the set of rational numbers.	This follows from ________.

(c) some rational numbers are integers.
(d) all rational numbers are integers.
(e) the sets of rational numbers and integers have the same cardinality.

13. In Table Test 2-2, there are two blanks in the **Reasons** column. What mathematical expression should go in the first blank (the one in the fourth line)?

    (a) k/k
    (b) $k/0$
    (c) $1/k$
    (d) $0/k$
    (e) None of the above

14. What words should go in the second blank (the one in the fifth line) of Table Test 2-2?

    (a) the quotient-of-integers axiom.
    (b) the commutative property for division.
    (c) the transitive property.
    (d) the definition of rational number.
    (e) the definition of integer.

15. Suppose you are reading a theoretical text, and you come across the following principle of arithmetic:

$$(\forall x)(\forall y)\ [(x \in \boldsymbol{Q})\ \&\ (y \in \boldsymbol{Q})] \Rightarrow \{(xy = 0) \Leftrightarrow [(x = 0) \vee (y = 0)]\}$$

What does this mean in plain English?

(a) For some rational numbers x and for some rational numbers y, the product xy equals zero if and only if x equals zero or y equals zero, or both.
(b) For all rational numbers x and for all rational numbers y, if the product xy equals zero, then x equals zero or y equals zero, or both.
(c) For all rational numbers x and for all rational numbers y, if x equals zero or y equals zero, or both, then the product xy equals zero.
(d) For all rational numbers x and for all rational numbers y, the product xy equals zero if and only if x equals zero or y equals zero, or both.
(e) For some rational numbers x and for some rational numbers y, if the product xy equals zero, then x equals zero or y equals zero, or both.

16. One of the most famous axioms in geometry states that if you have a line L and a point P not on line L, then there exists one and only one line M through point P that is parallel to line L. But some mathematicians investigated the consequences of denying this axiom. It turns out that the axiom does not hold true

(a) on flat surfaces.
(b) when all lines are perfectly straight.
(c) on the surface of a sphere.
(d) in Euclidean 3-dimensional space.
(e) in any case; it was a huge mistake to begin with.

17. In Fig. Test 2-3, suppose the arcs (shown as dashed lines) are centered at points P and Q, and that both arcs have been drawn with a drafting compass set for the same radius. In this case, point T

(a) lies at the center of a square whose vertices are points P, Q, R, and S.
(b) lies on the hypotenuse of a right triangle whose vertices are points R, T, and Q.
(c) lies at one of the vertices of a right triangle whose other two vertices are points R and Q.
(d) lies at one of the vertices of an isosceles triangle whose other two vertices are points S and P.
(e) lies at the center of a circle passing through points P, Q, R, and S.

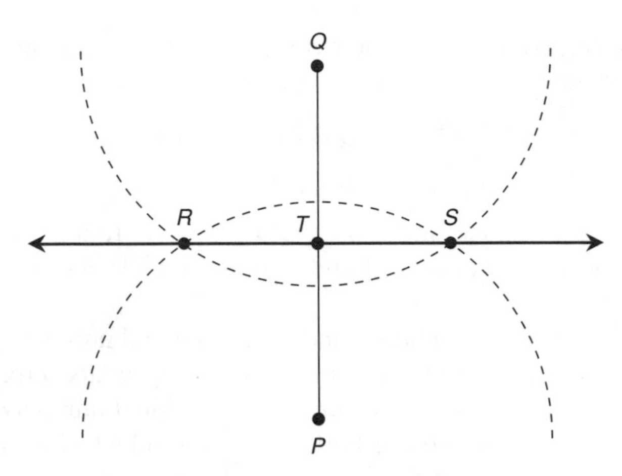

Fig. Test 2-3. Illustration for Part Two Test Question 17.

18. Imagine that you invent a new mathematical relation among geometrical objects called "yotto-congruence." You symbolize this relation by means of a yen symbol (¥). You are able to prove that for any three geometric figures Q, R, and S, the following three properties hold true:

$$Q \yen Q$$
$$(Q \yen R) \Rightarrow (R \yen Q)$$
$$[(Q \yen R) \ \& \ (R \yen S)] \Rightarrow (Q \yen S)$$

From this, by definition, yotto-congruence is

(a) a reflexive relation.
(b) a transitive relation.
(c) a commutative relation.
(d) an existential relation.
(e) an equivalence relation.

19. Which of the following statements (a), (b), (c), or (d), if any, is an accurate verbal description of the Pythagorean Theorem?

(a) Given a right triangle, the square of the length of the hypotenuse is equal to the sum of the squares of the lengths of the other two sides, if all lengths are expressed in the same units.
(b) Given a right triangle, the length of the hypotenuse is equal to the sum of the lengths of the other two sides, if all lengths are expressed in the same units.

(c) Given a right triangle, the length of the hypotenuse is equal to the sum of the squares of the lengths of the other two sides, if all lengths are expressed in the same units.
(d) Given a right triangle, the square of the length of the hypotenuse is equal to the sum of the lengths of the other two sides, if all lengths are expressed in the same units.
(e) None of the above statements (a), (b), (c), or (d) is an accurate verbal description of the Pythagorean Theorem.

20. Drawings can be used to help illustrate the meaning of a definition. But a drawing should never
(a) be published along with a definition.
(b) be made with any tools other than a straight edge and compass.
(c) be a necessary part of a definition.
(d) violate the formal rules of geometric construction.
(e) accompany a proof in set theory or logic.

21. Suppose you are told that if the sum of the digits in an integer n is equal to 9, then the integer n is divisible by 9 without a remainder. Which, if any, of the following statements is logically equivalent to this?
(a) If an integer n is divisible by 9 without a remainder, then the sum of the digits in the integer n is equal to 9.
(b) An integer n is divisible by 9 without a remainder if and only if the sum of the digits in the integer n is equal to 9.
(c) If an integer n is not divisible by 9 without a remainder, then the sum of the digits in the integer n is not equal to 9.
(d) An integer n is not divisible by 9 without a remainder if and only if the sum of the digits in the integer n is not equal to 9.
(e) None of the above

22. Any real number x is equal to itself. This is an example of
(a) the commutative property.
(b) the associative property.
(c) the transitive property.
(d) the symmetric property.
(e) None of the above

23. Suppose you want to prove that a particular positive integer q is composite. Which of the following can you do to accomplish this?
(a) Prove that q/q is a positive integer.
(b) Prove that $q/1$ is a positive integer.

(c) Prove that q^2 is an odd positive integer.
(d) Prove that $2q$ is an even positive integer.
(e) None of the above are sufficient to prove that q is composite.

24. Suppose P, Q, and R are the vertices of an equilateral triangle. Then it is possible to prove that P, Q, and R
(a) coincide.
(b) are collinear.
(c) are mutually equidistant from each other.
(d) do not all lie in the same plane.
(e) define a unique line segment.

25. Suppose you want to prove that a given number x is rational. Which of the following can you do to accomplish this?
(a) Find two integers a and b such that $x = ab$, where $b \neq 0$.
(b) Find two integers a and b such that $x = a + b$, where $b \neq 0$.
(c) Find two integers a and b such that $x = a^b$, where $b \neq 0$.
(d) Find two integers a and b such that $x = a/b$, where $b \neq 0$.
(e) None of the above are sufficient to prove that x is rational.

26. Suppose you want to prove that the number 640,000,000,000 is a product of primes. How can you prove this without actually finding the primes?
(a) invoke the GCD (greatest common divisor) theorem.
(b) invoke the "weak version" of the Fundamental Theorem of Arithmetic.
(c) invoke *reductio ad absurdum*.
(d) invoke DeMorgan's law for multiplication.
(e) None of the above

27. No rational number is an irrational number, and no irrational number is a rational number. This is logically equivalent to saying that the sets of rational and irrational numbers are
(a) disjoint.
(b) empty.
(c) commutative.
(d) equivalent.
(e) reflexive.

28. In a formal geometric construction using a compass and straight edge, which of the following actions (a), (b), (c), or (d) is impossible or not allowed?

(a) Using a straight edge to draw a line segment connecting two specific points.
(b) Using a straight edge to denote a line through a specific point, but oriented arbitrarily (at random).
(c) Using a straight edge to denote a ray defined by two specific points, with the end of the ray at one point, and with the ray passing through the other point.
(d) Using a straight edge to denote a line defined by two specific points, such that the line passes through both points.
(e) All of the above actions (a), (b), (c), and (d) are possible, and all are allowed.

29. Imagine two real numbers, x and y. You aren't told specifically what they are, but you are told that it is not true that x is larger than y, and that it is also not true that x and y are equal. You conclude that x is smaller than y. You know this because it follows from

(a) the GCD (greatest common divisor) theorem.
(b) the "weak version" of the Fundamental Theorem of Arithmetic.
(c) *reductio ad absurdum*.
(d) DeMorgan's law for equality.
(e) None of the above

30. Suppose we want to prove that it is impossible to have a triangle with straight-line sides, and contained on a flat surface, with interior angles whose measures add up to 200°. What technique might we consider in attacking this problem?

(a) Mathematical induction.
(b) Proof by example.
(c) *Reductio ad absurdum*.
(d) Inductive reasoning.
(e) The probability fallacy.

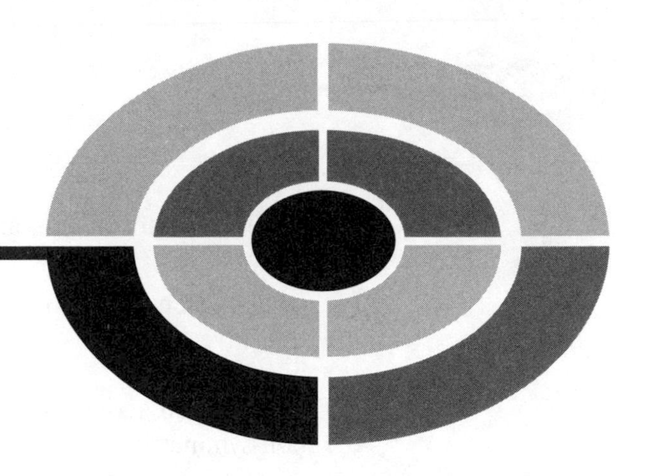

Final Exam

Do not refer to the text when taking this test. You may draw diagrams or use a calculator if necessary. A good score is at least 53 answers (75 percent or more) correct. Answers are in the back of the book. It's best to have a friend check your score the first time, so you won't memorize the answers if you want to take the test again.

1. The existential quantifier, symbolized ∃, can be translated into the words
   (a) "For all."
   (b) "For every."
   (c) "There exists."
   (d) "There does not exist."
   (e) "For one and only one."

2. Suppose you are given this axiom and told that it applies to a system of geometry you will be studying:
   - Let *L* be a straight line, and let *P* be some point not on *L*. Then it is not true that there is one and only one straight line *M*, in the plane defined by line *L* and point *P*, that passes through point *P* and that is parallel to line *L*.

From the statement of this axiom, you can be sure that the system of geometry is

(a) an equivalence relation.
(b) transitive.
(c) non-Euclidean.
(d) a DeMorgan geometry.
(e) non-symmetric.

3. Table Exam-1 is a truth table that denotes two variables (X and Y) along with several other propositions (P, Q, R, S, and T). What do the truth values in the column under proposition P represent?

   (a) X & Y
   (b) $X \vee Y$
   (c) $X \Rightarrow Y$
   (d) $X \Leftrightarrow Y$
   (e) $\neg X$

4. In Table Exam-1, what do the truth values in the column under proposition Q represent?

   (a) X & Y
   (b) $X \vee Y$
   (c) $X \Rightarrow Y$
   (d) $X \Leftrightarrow Y$
   (e) $\neg X$

Table Exam-1. Truth table for Final Exam Questions 3 through 7.

X	Y	P	Q	R	S	T
F	F	T	F	T	F	T
F	T	T	T	F	F	T
T	F	F	T	F	F	F
T	T	F	T	T	T	T

5. In Table Exam-1, what do the truth values in the column under proposition R represent?
   (a) X & Y
   (b) X ∨ Y
   (c) X ⇒ Y
   (d) X ⇔ Y
   (e) ¬X

6. In Table Exam-1, what do the truth values in the column under proposition S represent?
   (a) X & Y
   (b) X ∨ Y
   (c) X ⇒ Y
   (d) X ⇔ Y
   (e) ¬X

7. In Table Exam-1, what do the truth values in the column under proposition T represent?
   (a) X & Y
   (b) X ∨ Y
   (c) X ⇒ Y
   (d) X ⇔ Y
   (e) ¬X

8. Suppose that you want to prove that a certain property holds true for all the negative integers; that is, for the following set:

   $$\mathbf{Z}_{-} = \{-1, -2, -3, -4, -5, \ldots\}$$

   Which logical technique suggests itself in a case like this?
   (a) *reductio ad absurdum.*
   (b) the commutative law for integers.
   (c) DeMorgan's law for integers.
   (d) mathematical induction.
   (e) the law of implication reversal.

9. Suppose that it is the middle of the winter, and you're listening to the weather forecast on the radio. The disc jockey says, "It is probably snowing in Fairbanks, Alaska right now." By making this statement, the disc jockey

(a) commits DeMorgan's fallacy.
(b) commits the probability fallacy.
(c) makes an appeal to circumstance.
(d) commits improper use of context.
(e) is being perfectly valid and rigorous.

10. In mathematics, the term *rigorous* refers to
    (a) a logical process in which there exists at least one flaw or fallacy.
    (b) an axiom that does not apply to all cases for which it is supposedly intended, and which is therefore flawed.
    (c) a set of axioms and definitions that is inconsistent; that is, it ultimately contains a built-in contradiction.
    (d) a process or theory that is carried out or built up based entirely on valid logic.
    (e) inductive reasoning, in which something is shown to be true in most cases but not necessarily in all cases.

11. Refer to Fig. Exam-1. Suppose you want to prove that if ΔABC lies entirely in a flat plane, then the sum of the measures of angles x and y is equal to 90°. In order to do this proof easily, you would make use of
    (a) the fact that the sum of the measures of the interior angles of a plane triangle is always equal to 180°.
    (b) *reductio ad absurdum.*
    (c) DeMorgan's theorem for triangles.
    (d) the commutative principle for angles.
    (e) mathematical induction.

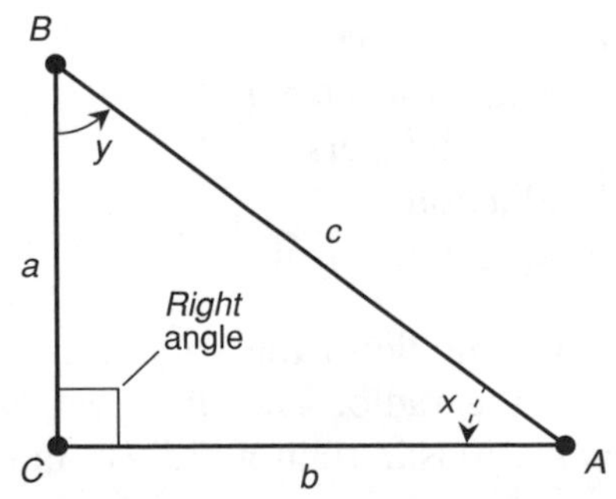

Fig. Exam-1. Illustration for Final Exam Questions 11 through 13.

12. Refer to Fig. Exam-1. Suppose you are given the following facts concerning ΔABC:
    - The length of side a is 15 centimeters ($a = 15$)
    - The length of side b is 20 centimeters ($b = 20$)
    - The length of side c is 23 centimeters ($c = 23$)

    Suppose you are told that the length measurements are exact, according to instruments designed to measure displacement to a tiny fraction of a centimeter. From these facts, along with the fact that in Euclidean geometry it is always true that $a^2 + b^2 = c^2$, you can prove that

    (a) this scenario is Euclidean.
    (b) this scenario is non-Euclidean.
    (c) none of the sides of ΔABC is straight.
    (d) ΔABC lies entirely in a flat plane.
    (e) ΔABC lies entirely on the surface of a sphere.

13. Refer to Fig. Exam-1. Suppose you are given the following facts concerning ΔABC:
    - The measure of angle x is 46° ($x = 46°$)
    - The measure of angle y is 44° ($y = 44°$)

    Suppose you are told that the angle measurements are exact, according to instruments designed to measure angles to a tiny fraction of a degree. From these facts, along with the fact that in Euclidean geometry it is always true that the sum of the measures of the interior angles is equal to 180°, you can prove that

    (a) this scenario is non-Euclidean.
    (b) at least one of the sides of ΔABC is not straight.
    (c) none of the sides of ΔABC is straight.
    (d) ΔABC lies entirely on the surface of a sphere.
    (e) None of the above

14. Fill in the blank to make the following sentence true and correct: "When proving a complicated theorem, it can help if we propose and prove one or more ________ first, using it, or them, to simplify the proof of the intended theorem."

    (a) definitions
    (b) axioms
    (c) corollaries
    (d) lemmas
    (e) implications

15. Consider two sets of things: the set of plants and the set of trees. You know that the set of trees is a proper subset of the set of plants. Let P represent the predicate "is a plant." Let T represent the predicate "is a tree." Let x represent a logical variable. Which of the following statements is true?
    (a) $(\forall x)(\mathrm{P}x \Rightarrow \mathrm{T}x)$
    (b) $(\forall x)(\mathrm{P}x \Leftrightarrow \mathrm{T}x)$
    (c) $(\forall x)(\mathrm{P}x \,\&\, \mathrm{T}x)$
    (d) $(\exists x)(\mathrm{P}x \,\&\, \neg\mathrm{T}x)$
    (e) $(\exists x)(\neg\mathrm{P}x \,\&\, \mathrm{T}x)$

16. Refer to Table Exam-2. In the symbology of predicate logic, how is the sentence "Paul is a biology student" written?
    (a) Pb
    (b) pB
    (c) Bp
    (d) bP
    (e) This sentence cannot be symbolized using the data in the table.

17. Refer to Table Exam-2. In the symbology of predicate logic, how is the sentence "Carol endorsed three checks" written?
    (a) Ce
    (b) Ec3
    (c) Eccc
    (d) c3E
    (e) This sentence cannot be symbolized using the data in the table.

Table Exam-2. Table for Final Exam Questions 16 through 18.

Subject	Subject Symbol	Predicate	Predicate Symbol
Paul	p	wrote a check	W
Carol	c	will endorse a check	E
Sam	s	is a biology student	B
Molly	m	was a teacher	T

18. Refer to Table Exam-2. In the symbology of predicate logic, how is the sentence "Molly was a teacher" written?
    (a) *m*T
    (b) T*m*
    (c) T & *m*
    (d) Any of the above
    (e) This sentence cannot be symbolized using the data in the table.

19. Refer to Fig. Exam-2. From this graph, it is reasonable to suppose that
    (a) the relative frequency, intensity, or amount of phenomenon *X* is correlated (although in a negative way) with the relative frequency, intensity, or amount of phenomenon *Y*.
    (b) an increase in the relative frequency, intensity, or amount of phenomenon *X* causes a decrease in the relative frequency, intensity, or amount of phenomenon *Y*.
    (c) an increase in the relative frequency, intensity, or amount of phenomenon *Y* causes a decrease in the relative frequency, intensity, or amount of phenomenon *X*.
    (d) All three of the above (a), (b), and (c)
    (e) None of the above (a), (b), or (c)

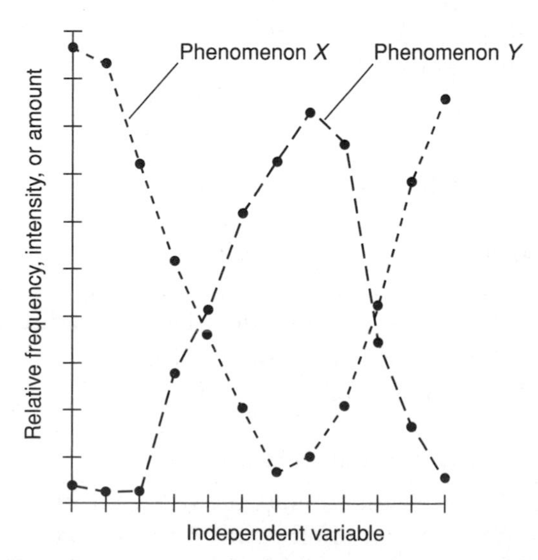

Fig. Exam-2. Illustration for Final Exam Questions 19 and 20.

20. Refer to Fig. Exam-2. What is meant by the term *independent variable* on the horizontal axis of the graph?
    (a) It is a factor that causes both phenomena X and Y.
    (b) It is a factor that is caused by phenomenon X or phenomenon Y, or both.
    (c) It is a factor that is caused by both phenomena X and Y.
    (d) It is a factor that is not influenced by either phenomenon X or phenomenon Y.
    (e) It is a factor that remains constant.

21. A straight line can be considered
    (a) zero-dimensional (0D).
    (b) one-dimensional (1D).
    (c) two-dimensional (2D).
    (d) three-dimensional (3D).
    (e) to have an undefined number of dimensions.

22. When two triangles have the same shape but are different in size, such that one can be "pasted down" right over the other if one of them is "magnified" to just the right extent, then the two triangles are
    (a) isosceles.
    (b) equilateral.
    (c) directly similar.
    (d) complementary.
    (e) None of the above

23. An elementary term is
    (a) a term that defies definition, and the meaning of which no one understands.
    (b) a term that is not formally defined, but the meaning of which is understood.
    (c) a term that requires an axiom in order to be defined.
    (d) a term that is an element of the set of integers.
    (e) a term that is not an element of any set.

24. How can you define the term *closed line segment*?
    (a) Let R and S be distinct points on a straight line X. The *closed line segment* RS is the set of all points on X between, but not including, points R and S.

(b) Let R and S be distinct points on a straight line X. The *closed line segment RS* is the set of all points on X between points R and S, including point R but not point S.
(c) Let R and S be distinct points on a straight line X. The *closed line segment RS* is the set of all points on X between points R and S, including point S but not point R.
(d) Let R and S be distinct points on a straight line X. The *closed line segment RS* is the set of all points on X between, and including, points R and S.
(e) Any of the above

25. Let A and B be two non-empty sets. Let x be a variable. Suppose that the following sentence is true:

$$(\forall x)\ (x \in B \Rightarrow x \in A)$$

From this, we can conclude that
(a) sets A and B are disjoint
(b) sets A and B are coincident
(c) sets A and B are non-disjoint and non-coincident
(d) set A is a subset of set B
(e) set B is a subset of set A

26. In propositional logic, the smallest logical element is
(a) a predicate.
(b) an existential quantifier.
(c) a universal quantifier.
(d) a sentence.
(e) an equivalence relation.

27. In a sentence containing a subject, a linking verb, and a subject complement, the predicate consists of
(a) the linking verb and the subject complement.
(b) the linking verb only.
(c) the subject complement only.
(d) the subject and the linking verb.
(e) the subject and the subject complement.

28. Once a proposition has been proved within the framework of a mathematical system, that proposition becomes
(a) an axiom.
(b) a theorem.

(c) an equivalence relation.
(d) a contradiction.
(e) an antecedent.

29. A corollary is
    (a) a theorem that arises as a secondary result from a significant theorem.
    (b) a theorem that is used as part of the proof of a more important theorem.
    (c) a theorem that ultimately results in a contradiction.
    (d) a theorem proved using *reductio ad absurdum.*
    (e) a theorem proved using mathematical induction.

30. Refer to Fig. Exam-3. Suppose that both of these triangles lie entirely in a single plane, and you want to prove that the triangles are inversely congruent. You can do this quickly and in a straightforward manner based on
    (a) the angle-angle-angle (AAA) axiom.
    (b) the angle-side-angle (ASA) axiom.
    (c) the side-angle-angle (SAA) axiom.
    (d) the side-side-side (SSS) axiom.
    (e) None of the above axioms, because these two plane triangles are not inversely congruent.

31. In a geometric construction, the straight edge should
    (a) be as long as possible.
    (b) be graduated in units such as centimeters or inches.
    (c) not be graduated in units such as centimeters or inches.
    (d) be made of clear material to allow optimum visibility.
    (e) be one edge of a drafting triangle.

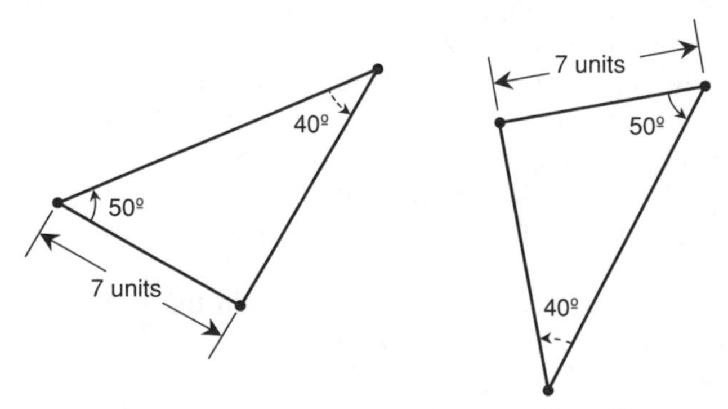

Fig. Exam-3. Illustration for Final Exam Questions 30.

32. In a geometric construction, which of the following actions is allowed?
    (a) Drawing a straight line 10 inches long.
    (b) Drawing a circle with a radius of 5 centimeters.
    (c) Drawing an arc with a measure of 37°.
    (d) Drawing a right angle as part of the process of bisecting a line segment.
    (e) None of the above (a), (b), (c), or (d) are allowed.

33. The universal quantifier, symbolized $\forall$, can be translated into the words
    (a) "For every."
    (b) "There exists."
    (c) "There does not exist."
    (d) "For one and only one."
    (e) Nothing! There is no such thing as a universal quantifier!

34. According to some scientists, the ice ages (in which vast regions of the earth's land mass were covered by glaciers) took place because the sun got dimmer, thereby allowing the whole planet to cool off. What can be said with rigorous mathematical certainty about this?
    (a) It is impossible, because no one with scientific instruments was there.
    (b) It is possible, because some people think things happened that way.
    (c) It is likely, because a lot of people think things happened that way.
    (d) It can be proven by inductive reasoning.
    (e) Things either happened that way, or else they did not.

35. Refer to Fig. Exam-4. Suppose that both of these triangles lie entirely in a single plane, and you want to prove that the triangles are directly similar. You can do this quickly and in a straightforward manner based on

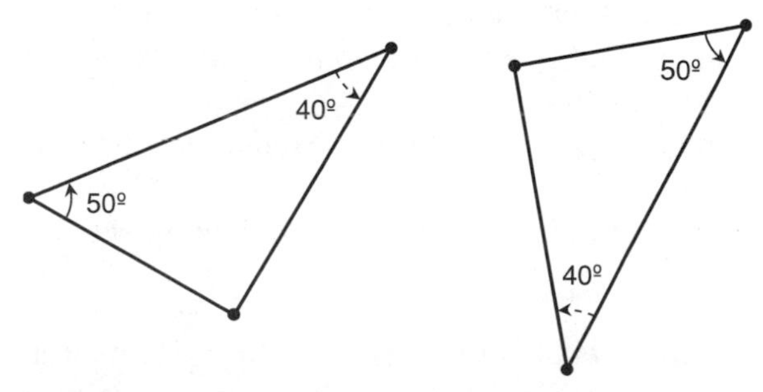

Fig. Exam-4. Illustration for Final Exam Questions 35.

(a) the fact that the sum of the measures of the interior angles of a plane triangle is equal to 180°, so we can deduce the measures of all three angles of both triangles; then we can invoke the definition of direct similarity.
(b) the fact that the sum of the measures of the interior angles of a plane triangle is equal to 180°, so we can deduce the measures of all three angles of both triangles; then we can invoke the angle-angle-angle (AAA) axiom.
(c) the fact that the sum of the measures of the interior angles of a plane triangle is equal to 180°, so we can deduce the measures of all three angles of both triangles as well as the length of one of its sides; then we can invoke the side-angle-angle (SAA) axiom or the angle-side-angle (ASA) axiom.
(d) the fact that the sum of the measures of the interior angles of a plane triangle is equal to 180°, so we can deduce the lengths of all three sides of both triangles; then we can invoke the side-side-side (SSS) axiom.
(e) None of the above, because there isn't enough information given in the figure to prove anything in particular about the two triangles.

36. Suppose there are two phenomena, symbolized F and G. Suppose F and G are correlated. This means that

(a) F causes G.
(b) If G is true, then F is true.
(c) If F is true, then G is true.
(d) F is logically equivalent to G.
(e) None of the above

37. Refer to Table Exam-3. What does this prove?

(a) An element in the intersection of two sets is not in their union.
(b) An element in the union of two sets is not in their intersection.
(c) An element that is not in the union of two sets is not in their intersection.
(d) An element that is in both of two disjoint sets is not in their intersection.
(e) An element that is in neither of two disjoint sets is in their intersection.

38. In Table Exam-3, what is the reason for the statement in line 5?

(a) This follows from the commutative law of disjunction.

Table Exam-3. Table for Final Exam Questions 37 and 38.

Statements	Reasons
Let A and B be non-empty sets.	We will use these in the proof.
Let c be a constant.	We will use this in the proof.
Assume $c \notin A \cup B$.	This is our initial assumption.
$\neg[(c \in A) \vee (c \in B)]$.	This follows from the definition of set union.
$\neg(c \in A)$ & $\neg(c \in B)$.	You'll get a chance to provide this reason in Exam Question 38.
$(c \notin A)$ & $(c \notin B)$.	This is simply another way of stating the previous line.
$c \notin A \cap B$.	This follows from the definition of set intersection.

(b) This follows from the associative law of conjunction.
(c) This follows from the distributive law of disjunction.
(d) This follows from the law of implication reversal.
(e) This follows from one of DeMorgan's principles.

39. According to the well-ordering axiom, every non-empty set of positive integers contains
    (a) a largest element.
    (b) an element that is equal to the mathematical average of all the elements.
    (c) a prime number.
    (d) a composite number.
    (e) a smallest element.

40. Using one of Euclid's postulates, it is easy to prove that all right angles
    (a) are equal in measure.
    (b) are equivalent to ¼ revolution.
    (c) have measures of 90°.
    (d) have the same measure as the angle at which two perpendicular lines intersect.
    (e) have all four of the above properties (a), (b), (c), and (d).

41. The *absurdum* quantifier, symbolized ∀, can be translated into the words
    (a) "For every."
    (b) "There exists."
    (c) "There does not exist."
    (d) "For one and only one."
    (e) Nothing! There is no such thing as an *absurdum* quantifier!

42. Refer to Fig. Exam-5. This is an example of
    (a) a logic flowchart.
    (b) an implication diagram.
    (c) a geometric construction of the bisection of an angle.
    (d) a Venn diagram.
    (e) a DeMorgan diagram.

43. Refer to Fig. Exam-5. The shaded region represents
    (a) set A only.
    (b) set B only.
    (c) the intersection of sets A and B.
    (d) the union of sets A and B.
    (e) None of the above

44. Refer to Fig. Exam-5. The region that is not shaded represents
    (a) set A only.
    (b) set B only.
    (c) the intersection of sets A and B.
    (d) the union of sets A and B.
    (e) None of the above

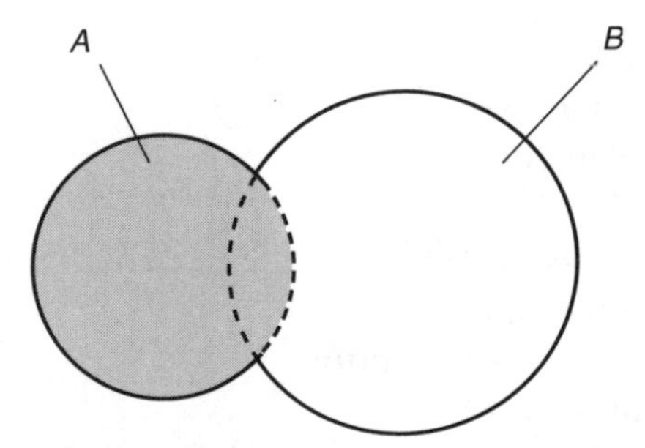

Fig. Exam-5. Illustration for Final Exam Questions 42 through 44.

45. Conjunction is an operation that can best be represented by the word or phrase
    (a) "or."
    (b) "logically implies."
    (c) "and."
    (d) "not."
    (e) "if and only if."

46. Disjunction is an operation that can best be represented by the word or phrase
    (a) "or."
    (b) "logically implies."
    (c) "and."
    (d) "not."
    (e) "if and only if."

47. A straight angle is the equivalent of
    (a) ¼ revolution.
    (b) ½ revolution.
    (c) ¾ revolution.
    (d) a complete revolution.
    (e) any integral multiple of ½ revolution.

48. Suppose you want to prove that all of the positive integers have a certain property. What method suggests itself here?
    (a) Mathematical induction.
    (b) The law of implication reversal.
    (c) DeMorgan's principle.
    (d) *Reductio ad absurdum.*
    (e) Inductive reasoning.

49. Suppose we are able to prove that a particular number t cannot be expressed as the ratio of any two integers. By proving this, we have shown that t is
    (a) a rational number.
    (b) a composite number.
    (c) an irrational number.
    (d) a prime number.
    (e) not defined.

50. Suppose we show that a number w can be expressed as a ratio of integers with a nonzero denominator. This proves that w is
    (a) a rational number.
    (b) a composite number.
    (c) an irrational number.
    (d) a prime number.
    (e) not defined.

51. A mathematical proof that is carried out by demonstrating the truth or validity of a single example
    (a) is never acceptable, because it cannot be rigorous.
    (b) can work for some propositions that contain existential quantifiers.
    (c) always results in a contradiction.
    (d) gives rise to an infinite number of other examples.
    (e) can only be done using mathematical induction.

52. How can you define the term *coincident lines*?
    (a) Let A, B, C, and D be distinct points. Line AB, defined by points A and B, and line CD, defined by points C and D, are *coincident lines* if and only if points A, B, C, and D are coplanar.
    (b) Let A, B, C, and D be distinct points. Line AB, defined by points A and B, and line CD, defined by points C and D, are *coincident lines* if and only if points A, B, C, and D are coincident.
    (c) Let A, B, C, and D be distinct points. Line AB, defined by points A and B, and line CD, defined by points C and D, are *coincident lines* if and only if points A, B, C, and D are perpendicular to each other.
    (d) Let A, B, C, and D be distinct points. Line AB, defined by points A and B, and line CD, defined by points C and D, are *coincident lines* if and only if points A, B, C, and D lie at the vertices of a rectangle.
    (e) None of the above

53. Refer to Fig. Exam-6. Note the four shaded triangles. Their sides each have lengths s, t, and u, with right angles at the vertices connecting adjacent sides of lengths s and t. Any two of these four shaded triangles can be proven directly congruent in a single step using either
    (a) the side-side-side (SSS) axiom or the angle-angle-angle (AAA) axiom.
    (b) the side-side-side (SSS) axiom or the side-angle-angle (SAA) axiom.
    (c) the side-side-side (SSS) axiom or the angle-side-angle (ASA) axiom.
    (d) the side-side-side (SSS) axiom or the side-angle-side (SAS) axiom.

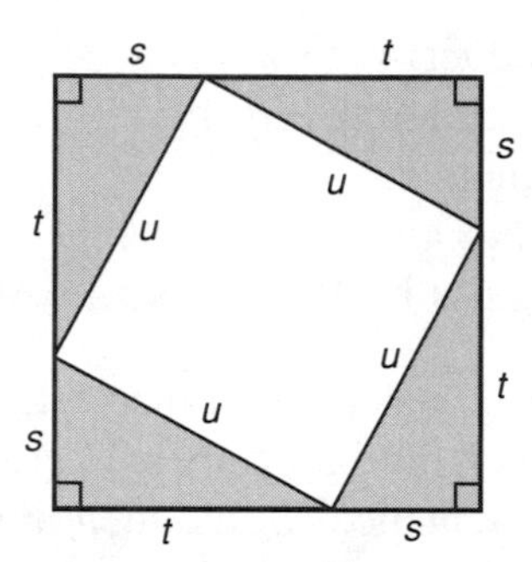

Fig. Exam-6. Illustration for Final Exam Questions 53.

(e) the side-angle-angle (SAA) axiom or the angle-angle-angle (AAA) axiom.

54. Suppose we have a huge positive integer. Call it *n*. We suspect that *n* is composite, but we want to prove it. How can we do this?
    (a) Use a computer to test every positive integer *k* such that $1 < k < n$, and see if any of them is equal to exactly $n/2$. If any of them is, then *n* is not composite. If none of them is, then *n* is composite.
    (b) Use a computer to test every positive integer *k* such that $k > n$, and see if any of them is equal to exactly $2n$. If any of them is, then *n* is composite. If none of them is, then *n* is not composite.
    (c) Use a computer in an attempt to find a set of primes $\{p_1, p_2, p_3, \ldots, p_k\}$, where *k* is some positive integer larger than 1, such that that $n = p_1 \times p_2 \times p_3 \times \ldots \times p_k$. If such a set can be found, then *n* is composite. If no such set can be found, then *n* is not composite.
    (d) Use a computer in an attempt to find a set of primes $\{p_1, p_2, p_3, \ldots, p_k\}$, where *k* is some positive integer larger than 1, such that that $n = p_1 + p_2 + p_3 + \ldots + p_k$. If such a set can be found, then *n* is not composite. If no such set can be found, then *n* is composite.
    (e) Use a computer in an attempt to prove that no positive integer smaller than *n* can be composite, by testing each and every one of them. If such a proof can be executed, then *n* is composite. Otherwise, *n* is not composite.

55. Suppose you want to prove that there is no such thing as a largest positive integer that is a product of prime numbers. What method suggests itself here?

(a) Mathematical induction.
(b) The law of implication reversal.
(c) DeMorgan's principle.
(d) *Reductio ad absurdum.*
(e) No method suggests itself, because the proposition is not true.

56. Let the predicate M represent "is a man." Let the predicate F represent "like (or likes) to watch football games." Let x be a logical variable. How would you write the sentence "Some men like to watch football games" in predicate logic symbology?

(a) $\exists M \;\&\; \exists F$
(b) $\exists M \Rightarrow F$
(c) $Mx \Rightarrow Fx$
(d) $(\exists x)\ (Mx \;\&\; Fx)$
(e) $(\exists x)\ (M \Rightarrow F)$

57. The if/then operation in propositional logic can be represented by the word or phrase

(a) "or."
(b) "logically implies."
(c) "and."
(d) "not."
(e) "if and only if."

58. When two triangles have exactly the same size and shape, so that one can be "pasted down" on top of the other without flipping either of them over (although rotation is allowed), the two triangles are

(a) isosceles.
(b) equilateral.
(c) inversely similar.
(d) complementary.
(e) None of the above

59. Let A and B be two non-empty sets. Let x be a variable. Suppose that the following sentence is true:

$$(\forall x)\ (x \in A \Rightarrow x \notin B)$$

From this, we can conclude that

(a) sets A and B are disjoint
(b) sets A and B are coincident

(c) sets *A* and *B* are non-disjoint and non-coincident
(d) set *A* is a subset of set *B*
(e) set *B* is a subset of set *A*

60. Suppose you want to prove that if a number is not an integer, then it cannot be a rational number. What method suggests itself here?

(a) Mathematical induction.
(b) The law of implication reversal.
(c) DeMorgan's principle.
(d) *Reductio ad absurdum.*
(e) No method suggests itself, because the proposition is not true.

61. Consider the following statement in propositional logic:

$$[(X \vee Y) \vee Z] \Leftrightarrow [X \vee (Y \vee Z)]$$

This is an expression of

(a) the associative law for disjunction.
(b) DeMorgan's law for conjunction.
(c) the law of implication reversal.
(d) the law of logical equivalence.
(e) *reductio ad absurdum.*

62. Suppose you want to prove the proposition $(\exists x)$ Px & Qx. Let k be a constant, and an element of the set for which the variable x is defined. In order to prove the proposition using the constant k, the minimum that we must do is show the truth of the statement

(a) Pk $\vee$ Qk.
(b) Pk.
(c) Qk.
(d) At least one of the statements (a), (b), or (c)
(e) Both of the statements (b) and (c)

63. Suppose Jim owns the only dry cleaning company in the town of Blissville. It is a one-person operation; he is the only employee. Jim, like every other adult in Blissville, owns a business suit. Jim cleans the business suits for all the adults, but only those adults, in Blissville who don't clean their own business suits. What can be "proven" about Jim?

(a) If Jim cleans his own business suit, then he does not.
(b) If Jim does not clean his own business suit, then he does.
(c) Jim does not exist.

(d) This scenario is a paradox.
(e) All of the above

64. Logical equivalence can be represented by the word or phrase
    (a) "or."
    (b) "logically implies."
    (c) "and."
    (d) "not."
    (e) "if and only if."

65. Suppose you are building a mathematical theory, and you come up with a proof that a certain statement is true. Later, you come up with a proof that the negation of the same statement is true. Which of the following cannot possibly be the case?
    (a) This always happens sooner or later in the process of mathematical theory-building, and it's nothing to worry about.
    (b) Your set of axioms is inconsistent.
    (c) One or more of the proofs you have done up to this point contains a flaw.
    (d) Your entire theory is flawed because it contains a contradiction.
    (e) You should consider eliminating one or more of your axioms, and starting the theory-building process all over again.

66. Refer to Fig. Exam-7. This shows the construction of a line segment between two specific points, P and Q. Using the straight edge alone (which in this case is one edge of a drafting triangle), we can, within the rules allowed for geometric constructions,

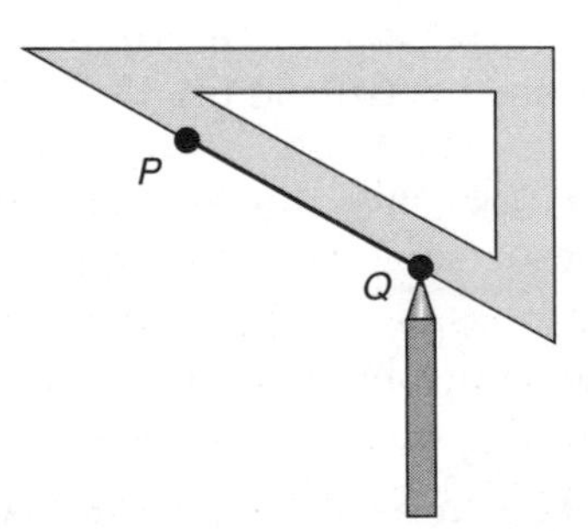

Fig. Exam-7. Illustration for Final Exam Questions 66 and 67.

(a) construct the midpoint of line segment PQ.
(b) extend line segment PQ in both directions to denote line PQ.
(c) construct a ray perpendicular to line segment PQ.
(d) construct a circle with radius equal to the length of line segment PQ.
(e) do none of the above operations (a), (b), (c), or (d).

67. Refer to Fig. Exam-7. This shows the construction of a line segment between two specific points, P and Q. Using the straight edge alone (which in this case is one edge of a drafting triangle), we can, within the rules allowed for geometric constructions,
    (a) extend line segment PQ past point P to denote the closed-ended ray QP.
    (b) construct an angle with a measure equal to the measure of any of the three angles at the vertices of the drafting triangle.
    (c) construct a line segment having twice the length of line segment PQ.
    (d) construct a line segment having any positive integral multiple of the length of line segment PQ.
    (e) do none of the above operations (a), (b), (c), or (d).

68. Suppose we have a huge positive integer. Call it n. We suspect that n is prime, but we want to prove it. How can we do this?
    (a) Use a computer to test every positive integer k such that $1 < k < n$, and see if any of them divides n without a remainder. If any of them does, then n is not prime. If none of them does, then n is prime.
    (b) Use a computer to test every positive integer k such that $1 < k < n$, and see if any of them divides n without a remainder. If any of them does, then n is prime. If none of them does, then n is not prime.
    (c) Use a computer in an attempt to find a set of primes $\{p_1, p_2, p_3, \ldots, p_k\}$, where k is some positive integer larger than 1, such that that $n = p_1 \times p_2 \times p_3 \times \ldots \times p_k$. If such a set can be found, then n is prime. If no such set can be found, then n is not prime.
    (d) Use a computer in an attempt to find a set of primes $\{p_1, p_2, p_3, \ldots, p_k\}$, where k is some positive integer larger than 1, such that that $n = p_1 + p_2 + p_3 + \ldots + p_k$. If such a set can be found, then n is prime. If no such set can be found, then n is not prime.
    (e) Use a computer in an attempt to prove that no positive integer larger than n can be prime, by testing each and every one of them. If such a proof can be executed, then n is prime. Otherwise, n is not prime.

69. Consider the following series of statements:

$(\forall x)\ (\mathrm{P}x \Rightarrow \mathrm{Q}x)$

$\neg \mathrm{Q}g$

$\neg \mathrm{P}g$

This is a symbolization of a proof by means of

(a) *reductio ad absurdum.*
(b) mathematical induction.
(c) DeMorgan's law for implication.
(d) the commutative law for implication.
(e) the law of implication reversal.

70. Which of the following (a), (b), or (c), if any, is an example of a subject/verb/object (SVO) sentence?

(a) Jim is a brilliant student.
(b) Paula is a soccer player.
(c) Ray was a math major.
(d) All of the above (a), (b), and (c) are SVO sentences.
(e) None of the above (a), (b), or (c) is an SVO sentence.

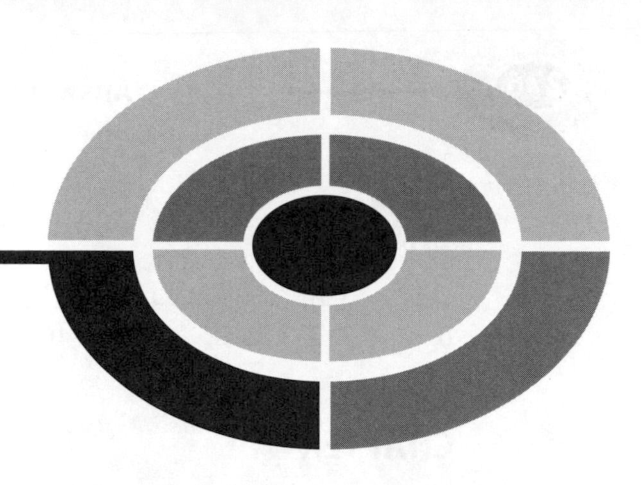

Answers to Quiz, Test, and Exam Questions

CHAPTER 1

1. b	2. a	3. d	4. c	5. a
6. a	7. b	8. b	9. d	10. d

CHAPTER 2

1. d	2. a	3. c	4. c	5. a
6. c	7. c	8. a	9. d	10. a

CHAPTER 3

1. c	2. d	3. c	4. a	5. b
6. a	7. d	8. d	9. a	10. a

CHAPTER 4

1. d	2. c	3. d	4. d	5. c
6. b	7. d	8. b	9. a	10. a

TEST: PART ONE

1. a	2. b	3. e	4. d	5. a
6. c	7. b	8. d	9. e	10. a
11. e	12. a	13. d	14. c	15. d
16. a	17. c	18. b	19. b	20. a
21. c	22. c	23. a	24. a	25. d
26. e	27. c	28. e	29. e	30. c
31. b	32. c	33. c	34. b	35. d
36. b	37. b	38. e	39. a	40. b

CHAPTER 5

1. b	2. a	3. c	4. b	5. d
6. d	7. c	8. b	9. d	10. b

CHAPTER 6

1. a	2. a	3. b	4. c	5. d
6. c	7. a	8. d	9. b	10. c

CHAPTER 7

1. d	2. b	3. c	4. a	5. d
6. b	7. b	8. d	9. b	10. c

TEST: PART TWO

1. e	2. c	3. a	4. c	5. e
6. c	7. d	8. b	9. a	10. c
11. b	12. b	13. e	14. d	15. d
16. c	17. c	18. e	19. a	20. c
21. c	22. e	23. e	24. c	25. d
26. b	27. a	28. e	29. e	30. c

FINAL EXAM

1. c	2. c	3. e	4. b	5. d
6. a	7. c	8. d	9. b	10. d
11. a	12. b	13. e	14. d	15. d
16. c	17. e	18. b	19. a	20. d
21. b	22. c	23. b	24. d	25. e
26. d	27. a	28. b	29. a	30. e
31. c	32. d	33. a	34. e	35. a
36. e	37. c	38. e	39. e	40. e
41. e	42. d	43. a	44. e	45. c
46. a	47. b	48. a	49. c	50. a
51. b	52. e	53. d	54. c	55. d
56. d	57. b	58. e	59. a	60. e
61. a	62. e	63. e	64. e	65. a
66. b	67. a	68. a	69. e	70. e

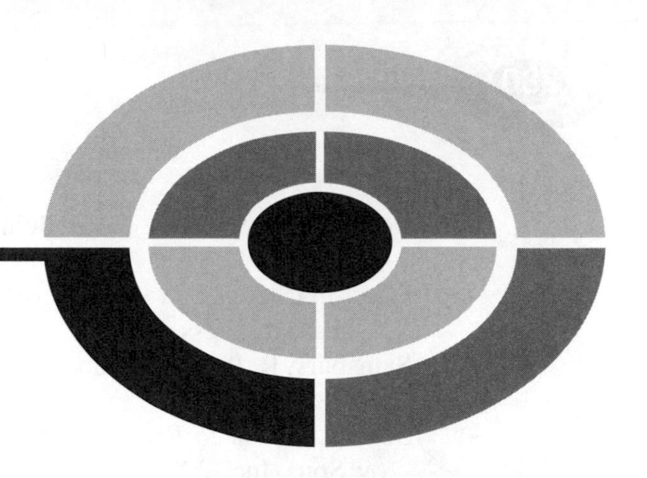

Suggested Additional References

Acheson, D. *1089 and All That: A Journey into Mathematics*. Oxford, England: Oxford University Press, 2002.

Berlinghoff, W. and F. Gouvêa. *Math through the Ages*. Farmington, ME: Oxton House Publishers, 2002.

Carnap, R. *Introduction to Symbolic Logic and Its Applications*. New York, NY: Dover Publications, 1958.

Courant, R., and H. Robbins. *What Is Mathematics?* 2nd ed. Oxford, England: Oxford University Press, 1996.

Cupillari, A. *The Nuts and Bolts of Proofs*, 2nd ed. San Diego, CA: Academic Press, 2001.

Dunham, W. *Journey through Genius: The Great Theorems of Mathematics*. New York, NY: John Wiley & Sons, Inc., 1990.

Euclid. *The Elements*. Santa Fe, NM: Green Lion Press, 2002.

Hardy, G. H. *A Course of Pure Mathematics*. Cambridge, England: Cambridge University Press, 1992.

Hardy, G. H. *A Mathematician's Apology*. Cambridge, England: Cambridge University Press, 1992.

Jacquette, D. *Symbolic Logic*. Belmont, CA: Wadsworth Publishing Company, 2001.

Priest, G. *Logic: A Very Short Introduction*. Oxford, England: Oxford University Press, 2000.

Sainsbury, R. *Paradoxes*, 2nd ed. Cambridge, England: Cambridge University Press, 1992.

Solow, D. *How to Read and Do Proofs,* 3rd ed. New York, NY: John Wiley & Sons, Inc., 2002.

Velleman, D. *How to Prove It: A Structured Approach*. Cambridge, England: Cambridge University Press, 1994.

INDEX

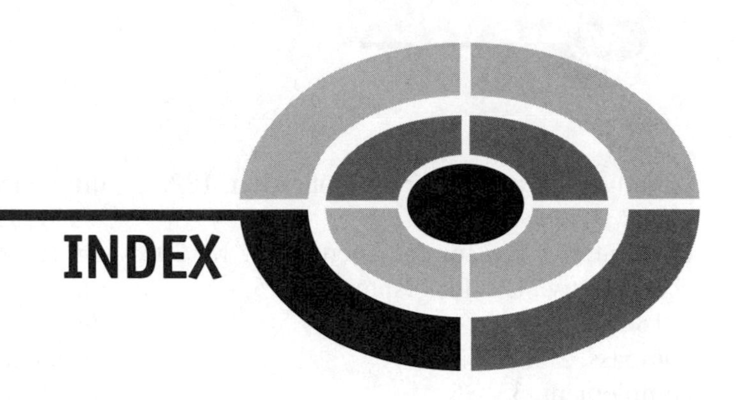

INDEX

287

ABOUT THE AUTHOR

Stan Gibilisco is one of McGraw-Hill's most prolific and popular authors. His clear, reader-friendly writing style makes his science, electronics, and mathematics books accessible to a wide audience. He is the author of *Teach Yourself Electricity and Electronics*, *Physics Demystified*, and *Statistics Demystified*, among more than two dozen other books and numerous magazine articles. Booklist named his *McGraw-Hill Encyclopedia of Personal Computing* one of the "Best References of 1996."